GRAVITATIONAL WAVES

GRAVITATIONAL WAVES

How Einstein's Spacetime Ripples Reveal the Secrets of the Universe

BRIAN CLEGG

ICON

Published in the UK and USA in 2018
by Icon Books Ltd, Omnibus Business Centre,
39–41 North Road, London N7 9DP
email: info@iconbooks.com
www.iconbooks.com

Sold in the UK, Europe and Asia
by Faber & Faber Ltd, Bloomsbury House,
74–77 Great Russell Street,
London WC1B 3DA or their agents

Distributed in the UK, Europe and Asia
by Grantham Book Services,
Trent Road, Grantham NG31 7XQ

Distributed in the USA
by Publishers Group West,
1700 Fourth Street, Berkeley, CA 94710

Distributed in Australia and New Zealand
by Allen & Unwin Pty Ltd,
PO Box 8500, 83 Alexander Street,
Crows Nest, NSW 2065

Distributed in South Africa
by Jonathan Ball, Office B4, The District,
41 Sir Lowry Road, Woodstock 7925

Distributed in India by Penguin Books India,
7th Floor, Infinity Tower – C, DLF Cyber City,
Gurgaon 122002, Haryana

Distributed in Canada by Publishers Group Canada,
76 Stafford Street, Unit 300
Toronto, Ontario M6J 2S1

ISBN: 978-178578-320-3

Typeset in Iowan by Marie Doherty

Printed and bound in Great Britain
by Clays Ltd, Elcograf S.p.A.

ABOUT THE AUTHOR

Brian Clegg's most recent books are *The Reality Frame* (Icon, 2017), *What Colour is the Sun?* (Icon, 2016) and *Are Numbers Real?* (St Martin's Press, 2017). He has also written *Big Data* in the Hot Science series. His *Dice World* and *A Brief History of Infinity* were both longlisted for the Royal Society Prize for Science Books. Brian has written for numerous publications including *The Wall Street Journal*, *Nature*, *BBC Focus*, *Physics World*, *The Times*, *The Observer*, *Good Housekeeping* and *Playboy*. He is editor of popularscience.co.uk and blogs at brianclegg.blogspot.com.

www.brianclegg.net

For Gillian, Chelsea and Rebecca

ACKNOWLEDGEMENTS

My thanks to the team at Icon Books involved in producing this series, notably Duncan Heath, Simon Flynn, Robert Sharman and Andrew Furlow. I've had great support in writing this from LIGO – notably Michael Landry – the ESA and the Max Planck Institute. A particular thank you to Kip Thorne for his wonderfully informative Andrew Chamblin Memorial Lecture at the Department of Applied Mathematics and Theoretical Physics in Cambridge.

CONTENTS

TIMELINE

1846 – Faraday gives his 'Thoughts on Ray-Vibrations' lecture in which he speculates in passing that gravitation may involve some kind of travelling wave

1916 – Einstein writes his first paper on gravitational waves as an outcome of his general theory of relativity

1918 – Einstein issues a second paper correcting an error in the first

1922 – Arthur Eddington suggests Einstein's waves are imaginary, created by the way the mathematics was used

1936 – Einstein writes to Max Born, telling him that after work with Nathan Rosen, he no longer believes gravitational waves exist

1937 – After corrections to his paper with Rosen, Einstein revives gravitational waves, but believes they are so weak they will never be observed

1955 – Joseph Weber studies gravitational radiation with John Wheeler at the Institute for Advanced Study

1957 – Richard Feynman shows that gravitational waves could do work and hence be detected

1958–60 – Weber begins construction of resonant bars in an attempt to detect gravitational waves

1962 – Mikhail Gertsenshtein and Vladislav Pustovoit publish the first paper on the theoretical use of interferometers in gravitational wave detection

1967 – Rainer Weiss publishes first practical design for gravitational wave interferometers

1968 – Kip Thorne begins theoretical work on gravitational wave detection

1969 – Weber claims the first detection

1972 – A detector based on Weber's principles travels to the Moon on Apollo 17

1974 – Weber's findings are largely dismissed

1974 – Hulse and Taylor make indirect observation of gravitational waves due to their impact on orbital decay

1975 – Prototype interferometer with 3-metre arms built by German/UK team in Garching

1980 – Funding obtained for planning for a large-scale interferometer in the US

1981 – Prototype with 40-metre arms built at Caltech

1986 – The LIGO project gets its first unifying project director, Rochus Vogt

1991 – The first major LIGO funding obtained

1994 – Ground broken on building the Hanford observatory

1995 – Ground broken on the Livingston observatory

1995 – Work on the European GEO600 detector started

2002 – Initial LIGO goes live and runs to 2010 – no waves detected

2005 – Useable models of black hole interaction and the waves they would produce developed

2006 – GEO600 reaches expected sensitivity – no waves detected

2014 – BICEP2 project team claim to have detected gravitational waves in the cosmic microwave background radiation – later retracted

2015 – Advanced LIGO goes live

14 September 2015 – First gravitational wave detection

11 February 2016 – LIGO first gravitational wave detection made public

1 June 2017 – Detection announcements continue with the third detection with a high confidence

14 SEPTEMBER 2015 1

There are times when those working on a major science pro-ject receive public accolades. Typically, it's when the data from a live science run is released, and what has been an intense period of private work becomes public property, to be dissected by the researchers' scientific peers and cele-brated by the world's press. But on 14 September 2015, the huge team working on LIGO – more formally, the Laser Interferometer Gravitational Wave Observatory – had no such expectations. No one realised that 50 years of fruitless work was about to be rewarded in an unexpected fashion.

The immense LIGO experiment, covering two sprawl-ing sites in the US and supported by over 1,000 scientists working around the world, was undergoing an engineering run. This was routine technical testing before the gravi-tational wave observatory would go live a few days later. It was the eighth and final cycle of fine-tuning before things might get interesting. Yet around 7.00am Eastern Standard Time – midday in the UK – a first email was sent out to interested parties that signalled the beginning of the biggest change to astronomy since the introduction of telescopes.

On that day, our understanding of the universe took a leap forward.

The gravity detectives

To call LIGO an observatory appears to be a dramatic understatement, though that is exactly what it is. It comprises two vast sites over 3,000 kilometres (1,865 miles) apart. Each of the near-identical facilities, one based in Livingston, Louisiana and the other at Hanford, Washington state, is home to a pair of 4-kilometre (2.5-mile)-long tubes, 1.2 metres across, set at right angles to each other to form an L-shape. At each site, a laser passes along the pair of tubes to reflect off mirrors at the ends many times before the beams are brought together to form an optical interference pattern, a tiny set of fringes that gives a visible warning of incredibly small changes. The slightest variation in the length of the beams will produce a detectable effect, a change that was expected to happen in the presence of gravitational waves – ripples in the fabric of space and time that had been predicted by Albert Einstein back in 1916, but had never been detected.

The vast twin systems, including those 4-kilometre lengths of metal tubing, contain hardly any air. The presence of vibrating air molecules would scatter the laser beams, introducing 'noise' into the carefully monitored signal. Any sound vibrations and air currents buffeting the delicately suspended mirrors located at the ends of the tubes would equally destroy the detection process.

The Livingston detector site, Louisiana.
Caltech/MIT/LIGO Laboratory

The Hanford detector site, Washington state.
Caltech/MIT/LIGO Laboratory

The pressure inside those tubes is a remarkable trillionth of the atmospheric level. This took 40 days of gradual pumping to achieve, during which time the tubes were heated to over 150°C to expel as much gas as possible from the metal surfaces.

Just getting the tubes ready for that evacuation took immense care. Establishing delicate equipment in remote areas of the United States was not without its problems. The tubes are big enough, and took long enough to construct, for the local wildlife to take up residence. When a member of the team walked through the near-completed tubes at Livingston, he discovered that wasps, black widow spiders, mice and snakes had all moved in. And that meant acid-bearing urine leaving stains on the pristine stainless steel that would release vapour when the air was removed, requiring a major cleaning effort before that vacuum could be established. (That word 'stainless' in 'stainless steel' doesn't apply once acids are involved.)

Despite the intense vacuum within the operational tubes, their metal walls are just 3 millimetres thick – around the same as 50 sheets of standard A4 paper. Without the frequent reinforcing loops along the length of the tubes, the outside air pressure would crush them. The exterior of each tube is cased in concrete, not to resist the vacuum, but to cushion any outside impact. This is just as well, as a security truck collided with one of the tubes of the Hanford observatory at night. The driver suffered a broken arm, but the tube stayed intact. A damaged tube, allowing air at atmospheric pressure to pour in, would have been catastrophic. The resulting blast of air would have destroyed most of the detection system, causing many millions of dollars' worth of damage.

Because the arms extend for such a distance, their supports have to gradually increase in height along their length to cope with the curvature of the Earth. From one end to the other, there is more than a metre difference in height, needed to keep the tube perfectly straight. And this was just a small consideration in ensuring that the detectors can function properly. A far bigger issue was vibration.

To deal with the inevitable environmental vibrations, LIGO has a whole host of feedback systems, which monitor position and make tiny movements of the arms and components to compensate for changes. Positions are monitored 983,000 times a minute – once every 0.000061 seconds. The 'seismic isolation platforms' deal with the larger vibrations down to around 1 million times larger than the waves that LIGO has to detect. The remaining reduction is achieved by the remarkable suspension systems used to keep LIGO's mirrors from moving due to anything other than gravitational waves. These use four separate pendulum suspensions to dampen movement, dangling the mirrors from glass fibres just twice the thickness of a human hair, keeping the 40-kilogram 'test mass'* mirrors as stable as possible.

---

* The LIGO mirrors are known as 'test masses', as it is the effect of gravitational waves on their mass that is used in detection.

One of LIGO's test masses installed in its quad suspension system. The 40-kg test mass is suspended below the metal frame above by four silica glass fibres.

Caltech/MIT/LIGO Laboratory

Business as usual

During the engineering run in September 2015, all of LIGO's detection systems were in play, bringing the light beams into alignment and testing their functionality, with no thought of capturing a breakthrough observation of a gravitational wave. For over 50 years, scientists had been looking for the tiny distortions in space and time caused by a distant cosmic event that would add a new, powerful approach to the astronomer's armoury. They had never achieved a single result. Some even suggested that gravitational waves would be impossible to detect unless we could take the leap of building an observatory in space, as the tiniest local tremor was enough to confuse the incredibly delicate instruments. But for now, these worries were put to one side. No careers were at risk of yet another failed detection of these elusive waves on this run. It was simply a matter of ensuring that the technology behaved as it should.

However, just because this was an engineering run did not mean that the observatory was inactive. Unlike the dome of a traditional telescope, with shutters to prevent light coming in, there is nothing that can stop gravity getting through. Gravitational waves may be incredibly weak and difficult to detect, but nothing can hinder their progress across the universe. And the 14 September email told the members of the LIGO collaboration that an unexpected event had occurred.

We still tend to think of astronomers peering directly through telescopes – but even most traditional optical observatories are now automated, their observers located anywhere in the world. Detection in the case of gravitational wave observatories is not about seeing something in the sky, but about pinpointing subtle changes in a stream of data from the instruments. The origins of those first, few cautious emails on 14 September emphasise how far this kind of science has moved the work away from on-the-spot observers. There *are* people stationed at Hanford and Livingston, but they are mostly engineers, involved in the day-to-day running of the equipment. The earliest email comments from gravitational physicists originated in Hanover, Melbourne, Paris and Florida – the only one from the US (where most of the collaboration were still asleep), and that located far away from either detector.

There was a time when data like this would have to be searched by eye, giving a team of grad students sleepless nights as they worked through page after page of computer printouts, fighting their way along mind-numbing strings of numbers. Now, though, much of that initial sifting is done using computer algorithms. Some of these systems look for specific patterns that models predict will be produced by natural phenomena expected to generate gravitational waves. But the system that flagged up the event on 14 September, the cWB or 'coherent wave burst' pipeline, had no such preconceptions. It was merely looking for near-simultaneous bursts of activity recorded at the two facilities. And cWB flagged up that a strong wave pattern had been received at Livingston, followed by a remarkably similar burst of activity at Hanford, 7 milliseconds later.

Event alert

The first response to this alert was to check for hardware injections. During the engineering run, it is normal practice to produce artificial signals to test whether or not the detection systems at the two sites pick them up. But there were no known planned injections made in the period when the detection occurred.

That didn't mean that the event was certainly a real sighting of a gravitational wave. All kinds of checks still had to be made. After all, this wasn't supposed to be an observation run; many oddities could occur in the systems as they were being fine-tuned. For that matter, there was always the possibility that what was being recorded was a large-scale seismic vibration that had been picked up by both observatories – or even that two totally separate vibrations had just happened to occur at the same time. And the LIGO team were always aware of the unnerving possibility of a blind injection.

Although the scientists could confirm that there were no routine hardware injections planned, what was being detected could have been an artificial event that had been intentionally triggered without the scientists' knowledge. Such 'blind injections' play an important role in the operation of a complex set of instruments like LIGO. They make sure that those involved aren't allowing their own prejudices and desires to influence their interpretation of the results. After all, the observations they make are simply variations in an ever-changing data stream. How that data is interpreted is crucial, and because the scientists never know whether an event is artificial or real until they have fully analysed it, they can't be biased by wishful thinking.

Blind injections had already been used in previous runs of LIGO, raising hopes on two occasions when it appeared more and more likely that an observation of gravitational waves had been made, only to have those expectations crushed when the secrecy was lifted to reveal a fake event. In theory, blind injections weren't needed during an engineering run, as no one was intending to take the data seriously, so it seemed unlikely that this was the case on 14 September – but at this stage of the process, the scientists had no way of knowing for sure.

Over the next two days, excitement grew. The event seemed more and more likely not only to be a real one, but also to provide a very significant discovery. No one had expected gravitational waves to be obvious in the data stream, but these were clear, visible signals – so strong that, were this a real detection, they had both found gravitational waves and made the first-ever direct observation of black holes. In which case, the team was surely looking at a Nobel Prize. More than that, their work – which some still believed was pointless, because they thought that LIGO wasn't sensitive enough – would have been the first step into accessing the mysteries of the universe in a way that had never been possible before. They were genuinely at the frontier of science. Yet it would be many months before the details could be made public. Months during which the teams had to lie to colleagues and repeatedly try to quash the rumours that began to fly around the scientific community. The countdown to gravitational wave astronomy had begun.

Before we can follow how the LIGO discovery unfolded in detail, we've some groundwork to do. We'll see how Einstein predicted the existence of gravitational waves almost exactly a century before the discovery (a coincidence that would itself make some wonder if the whole thing was a hoax). We will uncover the controversy surrounding early attempts to detect gravitational waves using massive metal bars, explore the brave step into the dark that led to LIGO despite, rather than thanks to its management, and discover the remarkable cosmological events involving black holes and neutron stars that make gravitational wave detection possible.

First, though, we ought to sort out the most fundamental aspect of the whole discovery. This is all about gravitational waves – but what do we actually mean by a 'wave'?

WHAT IS A WAVE?

2

Everyone has come across waves – those ripples on the surface of the water that you see if you drop a stone into a pond, or the moving walls of foam and brine that come crashing onto a beach, sometimes with devastating force. But to get the hang of gravitational waves we need to take a step back from the specific examples and understand what's going on beneath.

The anatomy of a wave

At its most basic, a wave is a movement in a substance, where that movement changes cyclically as it travels forward. The most familiar form, like those waves on the beach, are known as transverse waves – their cyclic motion alternates at right angles to the direction the wave is travelling – up and down in the case of water waves, or side-to-side when we send a wave along a rope by flicking it.

A very simple transverse wave looks like this:

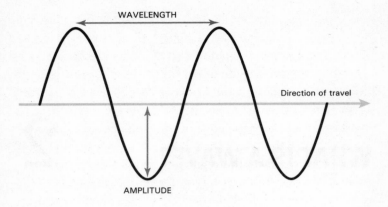

The wave is moving left to right, with the distance covered by a complete cycle of the wave known as the wavelength, and the number of such cycles that occur in a second being its frequency. A wave needs a 'medium' – stuff to actually do the waving, though in some cases, such as light, the nature of that medium is not immediately obvious. In the case of ocean waves, the medium is straightforwardly the water. A frequent misunderstanding is to assume that it is the water that moves forward – or more generally the medium – but actually it is the wave. Think of a Mexican wave travelling around a stadium (a Mexican wave is a transverse wave as the cycle of the motion is up and down, while the wave travels at right angles to that direction, round the stadium). The medium here is the mass of spectators who bob up and down. But they stay in their seat positions – they don't move forward around the stadium, only the wave does.

The other common form of wave is the longitudinal or compression wave. Perhaps the most familiar form of a longitudinal wave is sound, or the kind of wave you can send down a Slinky spring by giving its end a quick push. Here the cyclic motion isn't at right angles to the direction the wave is travelling, but back and forth in the same direction. Going at right angles wouldn't work for a sound wave, as it goes through the middle of the medium – the air. If it tried to go side-to-side, it would quickly lose its energy battling against the other air molecules. Transverse waves usually have to travel along the edge of the medium – for example, on the top of the water that the wave passes through. For a longitudinal wave, the regular cycle is in the same direction as the wave moves forward, not at right angles. The medium is repeatedly squashed up and relaxed like a concertina, so what travels through it is a pattern of compression and rarefaction.

A simple longitudinal wave looks like this:

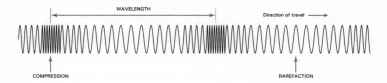

When you speak to someone, your vocal cords start a compression wave in the air that spreads out from your mouth until those compressions and rarefactions reach the listener's ear. There, they vibrate the hair-like structures in the ear, producing the sensation of hearing. But the link between you and the listener is the longitudinal waves that pass through the air.

Waves are very common occurrences in nature. Apart from waves on water and in sound, we find them, for example, travelling through the ground as a result of earthquakes. And you may well have been taught at school that light is a wave. In that example, though, we have to be a little more careful. Light certainly can act like a wave, but it's a little more tricky to pin down exactly what it *is*. It's worth getting an understanding of light, though, because it is the basis for almost all current astronomy – the discipline that gravitational waves have the potential to transform.

The model wave

For centuries, there have been arguments about the nature of light. We all are familiar with it, but it's intangible. It's difficult to pin down its nature. Some early scientists, such as Isaac Newton, argued that light was made up of a stream of particles. This would make sense of its value to astronomers. A stream of particles can flow across the vacuum of space to reach our eyes and telescopes. But a light wave shouldn't be able to cross totally empty space, because there's no medium to do the waving. Despite this restriction, others, notably Newton's contemporary Christiaan Huygens, thought that the light *was* a wave. Increasingly, over time, the wave theory of light became stronger, notably when it was observed that light displayed a common behaviour of waves called interference, which would prove hugely important in the gravitational wave story.

Imagine simultaneously dropping two stones, a few centimetres apart, into a still pond. The ripples – waves – that the stones create will head outwards from the two locations that the stones hit the water until those waves meet. When they do, there will be points on the surface of the water where both waves are rippling in the same direction (up or down) at the same time. Here, the waves will reinforce each other, becoming stronger than before. At other points on the surface, the waves will be rippling in opposite (vertical) directions at any point in time. Here the waves will cancel each other out, leaving relatively still patches of water. This effect, producing a distinctive pattern on the surface, is known as interference.

An interference pattern in water.
Shutterstock

In 1801, English scientist and polymath Thomas Young showed that light behaved exactly the same way as those ripples, apparently proving that it was a wave. When two beams of light were sent through nearby slits and the resultant beams overlapped, the result was an interference pattern of dark and light fringes. But there was a problem. As we have seen, unlike sound, light happily travels through the vacuum of space, where there is no medium for it to wave in. So how could it work?

Initially, the only possible explanation was that there was some kind of invisible, undetectable material that filled all space. This was known as the ether. But this would have to have been a very strange material – so insubstantial that we can't directly detect it, yet so rigid that light could travel through it for vast distances without losing energy to the floppiness. The remarkable Scottish physicist James Clerk Maxwell worked out in the early 1860s that light was an interaction between electricity and magnetism. And this meant that in principle, you could have an electric wave creating a magnetic wave, creating an electric wave and so on, hauling itself through empty space by its own bootstraps without any material medium required – it is the electromagnetic field that acts as the material.

This was the position in the early years of the twentieth century. However, quantum theory would blow a hole in comfortable Victorian assumptions. The work of Max Planck, Albert Einstein, Niels Bohr and others showed that light appeared to be both a wave *and* a stream of particles. Although it was convenient for many purposes to think of light as behaving like a wave, the particle idea explained more phenomena. As the great American physicist Richard Feynman would later put it: 'It is very important to know that light behaves like particles, especially for those of you who have gone to school, where you were probably told about light behaving like waves. I'm telling you the way it *does* behave – like particles.'

If you ask a physicist today what light is, they may well say that it is a travelling excitation in a quantum field. This is also a valuable analogy for light – though none of these descriptions provides us with its true nature. It's not that any are wrong – but each is just a kind of analogy. Light isn't a wave, or a stream of particles, or a disturbance in a quantum field, it's light. Each of these is a useful way of thinking about light in some circumstances. It's what's called in science a 'model' – not an actual description of reality, but a way of describing it that makes useful predictions.

So, we can say that light (sometimes) behaves like a wave, but not that it *is* a wave. This is in contrast to gravitational waves, which if they were to exist would actually be waves. Models are immensely useful, because we can rarely examine nature perfectly. We have to make do with what we can measure and detect and from that we build a model to describe how it behaves. But what is important for us in the story of gravitational waves is that light gives us a way to access distant parts of the universe. Unless it interacts with matter, light will carry on travelling indefinitely. There is light out there that has been travelling through space for billions of years. This makes it possible to examine different parts of the universe, and to see what it was like in the past. Light takes time to reach us, so the further it comes, the further back in time we are looking.

Light, though, has a number of limitations. It can be blocked by matter getting in the way and absorbing it. As well as big things like stars and planets, there is plenty of dust out in space that can prevent us from getting a good view. To make matters worse, we believe that the universe became transparent only when it was around 380,000 years old, 13.4 billion years ago. We can't look further back in time than this point. There's effectively a barrier for light waves. But gravitational waves would not be stopped by anything – they could help us fill in our astronomical gaps. *If* we could detect them.

To understand gravitational waves, it's important both to have a feel for the basics of waves and of what a model is, like the wave model of light, because the prediction of gravitational waves emerged from a model – specifically, Einstein's general theory of relativity, which is a brilliant mathematical model of the phenomenon we call gravity.

The gravity of the situation

Gravity is the least subtle of the forces of nature, though it is by far the weakest. If that sounds unlikely – gravity can seem pretty powerful – try picking up a pin with a magnet. The whole Earth is pulling the pin down with gravity, while the tiny magnet pulls it up with another of the forces of nature, electromagnetism. The magnet wins. We know there's something that keeps us stuck to the Earth and prevents us floating off into space. There's something that makes an object accelerate towards the Earth's surface when we release the object above the ground. There's something out there that makes the Moon orbit the Earth and the Earth orbit the Sun. That 'something' we call gravity.

2,400 years ago, the Ancient Greeks thought that everything around us was made up of four elements – earth, air, fire and water. Each of these had natural tendencies. Two – air and fire – had levity, which gave them a tendency to move away from the centre of the universe (or to put it another way, away from the Earth, which was considered to be that centre). By contrast, earth and water had gravity, which made them want to go towards the centre of the universe. 'Want' isn't quite the right word – the Greeks weren't suggesting the elements were conscious. Rather, they thought that they had a natural tendency, in the same way that trees grow upwards or water is wet.

It's not surprising that the Greeks had an interest in gravity. It is a force that is always with us, influencing our everyday life. What the Greeks didn't know was just how important gravity really is to us, not just to keep us firmly in place on the ground. There would be no Earth or stars or galaxies without gravity. It's the pull of gravity that over long periods of time causes gas and dust, scattered through space, to come together to form solar systems like our own. Not only did gravity cause the Sun and the Earth to form, it's also gravity that powers the Sun, compressing the hydrogen ions that largely make it up so much that they undergo the nuclear fusion reaction that generates the heat and light that keeps us alive.

The Ancient Greek way of thinking about gravity largely continued through to the 1500s, by which time their ideas were being challenged because they fitted so badly with what was actually observed. Because of the way gravity (and levity) was supposed to influence the four elements, it was assumed that the Earth should be made up of four concentric spheres. The innermost was earth, as this was most susceptible to gravity. Then came water, air and finally fire. It doesn't take a genius to spot that there's a problem with this model – all the solid stuff, the earth-based material, should be entirely surrounded by water. There would be no dry land.

To get around this, the model was tweaked by assuming that the sphere of earth was off-centre for some reason, so part of it protruded over the surface of the water. But once European ships had made their way to the New World and it became obvious that this was not part of the same land-mass as Europe, the gravity model could no longer hold up.

The discoveries from transatlantic voyages acted like an early scientific experiment – they tested the theory of the gravity/levity-driven elements and found it wanting. This proved fertile ground when someone with the imagination of Galileo began to think about the impact of gravity.

Gravity makes things fall

Just as we tend to think of Newton watching an apple fall, the archetypal image of Galileo is of the Italian scientist dropping balls off the Leaning Tower of Pisa to see how gravity influences objects of different weight. And like the Newton story, many believe this is a myth. Galileo was never shy about publicising his work, yet he failed to mention his Tower of Pisa experiment. The only reference we have to it was made by an assistant, writing in Galileo's old age, and the chances are it never happened. It's not really surprising – it wouldn't be easy to make measurements as balls came crashing to the ground. Instead, Galileo's work on gravity involved methods of falling that are restrained: the pendulum and the inclined plane.

One of the other great stories of Galileo, and there are many, is of him sitting in a service at the cathedral in Pisa, watching a huge chandelier swing from side to side. (Incidentally, this story was told by the same man as the dropped balls tale, so it too may be apocryphal.) Bored by the sermon, Galileo started to time the swinging of the chandelier, using his pulse as a timepiece. To his surprise he found that however big or small the swing of the pendulum, it took the same time to make its journey – an essential observation given the importance pendulums would have in clocks.

You might wonder what a pendulum has to do with falling and gravity. It's just a more complex sort of falling where there is a second force involved. The bob on the end of the pendulum falls towards the ground, but the other force, in this case the pull of the string linked to the mounting point, pulls it away from its fall. Galileo's more detailed study of gravity, where he discovered that the acceleration was independent of mass, involved rolling different weights of ball down a slope, minimising friction.

Again, there is a second force on the ball provided by the slope, but this can be removed from the calculation, and it was much easier to time and monitor balls on an inclined plane than just trying to follow them as they drop. The most dramatic direct demonstration of the uniform acceleration of gravity that Galileo established was direct, though. It was done on the Moon in 1971 by Apollo 15 astronaut David Scott, who simultaneously dropped a hammer and a feather. Without air resistance, both fell together (though significantly slower than they would on Earth).

Within twelve months of Galileo dying in 1642, Newton, the man we most associate with gravity, was born.

Mr Newton's magic model

Newton's story of a falling apple is slightly more believable than Galileo's with the Leaning Tower, because Newton did tell the story himself. Here's the historian William Stukeley's account of a conversation with Newton:

> After dinner, the weather being warm, we went into the garden, and drank thea [*sic*] under the shade of some apple

trees; only he and myself. Amidst other discourse, he told me, he was just in the same situation, as when formerly, the notion of gravitation came into his mind. Why should that apple always descend perpendicularly to the ground, thought he to himself; occasion'd by the fall of an apple, as he sat in a contemplative mood.

This would seem enough to give rise to the picture of Newton and the apple, but it's quite a leap to get from this to the idea of gravity keeping the orbiting planets in their place. Yet Stukeley's account goes on to describe the chain of thought that led from apple to universal gravitation. Of the apple, Stukeley observes:

Why should it not go sideways, or upwards? But constantly to the earths center? Assuredly the reason is, that the earth draws it. There must be a drawing power in matter. The sum of the drawing power in the matter of the earth must be in the earths center, not in any side of the earth. Therefore does this apple fall perpendicularly, or towards the center. If matter thus draws matter; it must be in proportion of its quantity. Therefore the apple draws the earth, as well as the earth draws the apple.

And you can see what may be the same apple tree – a 400-year-old example of the Flower of Kent variety that dates back to Newton's time – placed squarely in view of Newton's bedroom window at his old home, Woolsthorpe Manor in Lincolnshire. There's still a considerable amount of doubt about the story – Newton told it long after the event, and his work on gravity was mostly done many years after he left Lincolnshire – but it is entirely possible that thinking about an apple falling was his initial inspiration.

In his masterpiece, the *Principia* (1687), Newton laid out the mathematics that describes the force that keeps planets in orbit, or us on the Earth, though he used a singularly impenetrable style, with a lot of unnecessary geometry (unlike Galileo's still-enjoyable prose). Newton developed a very simple model of gravity, deciding that it was a characteristic of stuff – of matter. Stuff attracts other stuff. The more mass in the objects attracting each other, he argued, the greater the force with which they are pulled towards each other. And that force gets smaller with the square of the distance between the objects.

That's it, really. That's Newton's law of gravitation. That's all you need (with a spot of calculus to crunch the numbers) to work out how the Earth will orbit the Sun or how an apple will fall if you let it go at a certain height. The only trouble is that Newton had no idea how this gravity thing worked. His model was simply: 'There is an attraction between bits of stuff, and let's not bother about why.'

Newton underlined this lack of explanation by writing 'Hypotheses non fingo' (he was working in Latin), meaning 'I frame no hypotheses'. His attraction model caused quite a furore at the time, as the word 'attraction' was then used only in the sense of finding someone attractive, which seemed more than a little odd when applied to planets and falling bodies. Newton was mercilessly mocked for this. His force of gravity, this remote attraction, was called occult, in the sense of being hidden and mysterious.

Yet for all their bluster, his opponents had to admit that Newton's maths worked wonderfully. It predicted how things would fall and how heavenly bodies would orbit, tying together the heavens and the Earth with a single, universal law.

Despite claiming otherwise, Newton *did* have a hypothesis for how gravity worked, but thought it would unnecessarily muddy the water if he discussed it in his book. He suspected that there was a flow of pushy particles, moving in all directions through the universe. If one body (the Sun, say) blocked that flow of particles in a particular direction, another body in that direction (the Earth, say) would feel less pressure away from the Sun than towards – as a result it would be pushed towards the Sun, giving the impression of an attractive force. There were problems with this model – in its most basic form, it suggested the pull of gravity should depend on the size of things, not their mass. But it was the best Newton could come up with.

Newton's mathematical formulation of the force of gravity was brilliant, and it was sufficient to put men on the Moon. But it's not quite right – and it took the man who captured from Newton the 'most famous scientist ever' crown, Albert Einstein, to show why this was the case – and to dispose for ever of the problem of the mysterious remote attraction. In the early twentieth century, Einstein developed a model of gravity with a mathematical structure to describe how the force of gravity would behave, which would eventually lead to his prediction of the existence of gravitational waves.

EINSTEIN'S BABY 3

Albert Einstein started his voyage into understanding gravity, an approach that would be called the general theory of relativity, in Switzerland in 1907 where he was working before he got his first academic appointment. Famously, Einstein said that his 'happiest thought' occurred here: 'I was sitting in a chair in the Patent Office at Bern when all of a sudden a thought occurred to me. If a person falls freely he will not feel his own weight. I was startled.' By thinking of someone falling, for example in a plummeting lift, Einstein had realised that it was impossible to distinguish acceleration and the pull of gravity. And working through the mathematical implications of this made it clear that gravity was an effect that could be produced by a distortion of space and time.

The happiest thought

Imagine, for instance, you are in a spaceship without windows. You feel a steady force, pulling you towards the rear of the ship, just as if the ship were stationary, positioned back end downwards, on the surface of the Earth. Einstein pointed out that there was no way to distinguish between actually sitting still on the Earth, feeling the force of gravity, and being under a constant acceleration which would also produce that feeling of backwards pressure we get when a plane takes off. They were the same thing.

Now, when an object accelerates, it has a strange effect on things moving through its space. Imagine we're in a spaceship moving steadily through space and we throw a ball across the ship from side to side. It will move in a straight line, because both the ball and the ship are moving forwards at the same speed. But if the ship accelerates, the ball's path will become curved, as the ship moves ahead more quickly. The straight line path has become curved from the viewpoint of the ship. Similarly, Einstein reasoned, a straight line path in space would be curved by gravity.

The Sun and Earth warping space and time.
T. Pyle/Caltech/MIT/LIGO Lab

The model of gravity's warping of space that is often used is a rubber sheet. Imagine a flat sheet of rubber with a straight coloured line running through it. That line represents a beam of light, or the path of a planet moving through space. Now we put a bowling ball on the rubber. The sheet distorts as the ball produces an indentation. Look again at that line and it will now curve around the ball. So, for instance, even though the Earth is travelling in a straight line through space, space itself is curved, so the planet orbits the bowling ball of the Sun. It is a simple, but astonishing observation. Planets move in straight lines. There really is no force pulling them into an orbit. It's just that the space their straight line path runs through gets twisted.

That is remarkable, but it doesn't actually explain why Newton's apple fell. The apple isn't moving at all at the start of the process, so why would changing the shape of space make it move? The wonderful answer is that massive objects don't just warp space, they warp space and time. In Einstein's world, space and time are united into the single entity, spacetime. In principle we should think of spacetime as a four-dimensional object – but that's hard to envisage, so what we tend to do is to just use two space dimensions and one of time. (The third space dimension hasn't gone away, we just don't need to think of it.)

When we speak of warping space or time, what happens is that the axis is no longer straight, but starts to curve. (Imagine the space dimensions overleaf are on a piece of paper and we twist the paper.) Now let's go back to that apple steadily ticking through time – which is the same as moving up the time axis. Now we warp the time dimension. When we warp a spatial dimension, the result is move-ment in another spatial dimension. For example, if an ant is

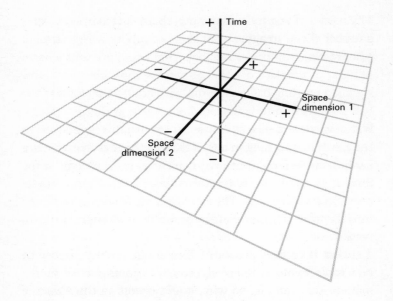

walking along on a vertical piece of paper and we twist that paper, it will now be moving horizontally. But there is only one time dimension, so the warp has nowhere to go but into space. Effectively part of the apple's progression through time becomes a movement through space. The apple begins to move because it's in a time warp.

There's another interesting consequence. As some of the effective movement in time is warped into movement in space, there is less movement in time left. Time slows down. This really works and has been widely tested. In fact, GPS satellites, which are just clocks pumping out a regular and very accurate time signal, have to be corrected for the way the weaker gravity in orbit means they run faster than they would on the Earth's surface. (They also have to be corrected for the way another of Einstein's discoveries, special relativity, slows them down, but that's another story, and gravity has the stronger effect.)

Waving the general theory

Einstein's model was much more sophisticated than Newton's, taking in all kinds of different factors that Newton had not considered, but it produced pretty much the same results in many cases. Rather than rely on a fuzzy model of 'things attract each other', Einstein used his powerful model which says that stuff warps space and time. Einstein's model also had the potential to make more complex predictions. The first proofs of Einstein's general theory of relativity involved relatively small variants on Newton's model. For example, Newton's version did not predict the orbit of Mercury exactly. For a while it was thought that the way Mercury's orbit shifts over time was being influenced by another unseen planet, given the name Vulcan, orbiting between Mercury and the Sun. But Einstein's theory matched observation exactly.

Similarly, Newton's theory under-predicted the amount by which the straight line path of light would be bent by massive bodies by a factor of two. But there were many other possibilities that could come out of Einstein's remarkable work, some of which are still being uncovered today.

The equations that Einstein developed, the so-called field equations of gravitation, are sufficiently complex that they can't be solved in a general-purpose, everything-you-ever-wanted-to-know way. Instead, they need to be solved for specific cases, such as an individual body – the first exact solution was for a non-rotating sphere and would eventually lead to the prediction of the existence of black holes. Another approach that was soon successful was producing solutions for a very simple model of the universe taken as a homogenous whole. And it was when Einstein was playing around with the implications of his general theory just a year after it was completed that he realised that one possibility was that there could be waves, not passing *through* space like the electromagnetic waves of light, but using as their medium the very fabric of space itself (or more precisely, waving in spacetime).

Gravitational waves have been described as being sounds that travel without a medium – but that's a misleading metaphor. A sound is a vibration of the molecules of matter. We are most familiar with air molecules providing that medium, though sound can travel just as easily through a liquid or solid. But gravitational waves are fundamentally different. Not only is a gravitational wave a side-to-side, lateral wave like light, not a longitudinal wave like sound, but it isn't a ripple through objects in spacetime; it passes through spacetime itself. When a gravitational wave passes by, spacetime squeezes and contracts. This oscillation influences matter, which exists in spacetime – but doesn't require matter to be able to travel. It's quite demeaning to gravitational waves to call their detection 'hearing the sound of black holes in space' as some have done. By comparison, sound is a trivial local effect. Gravitational waves make the universe itself vibrate. And to see how they might be produced and detected, we need to take a trip further into Einstein's universe.

When Newton first came up with his gravitational theory, it was generally assumed that the effect of gravity was instantaneous, whatever the distance between the bodies providing the attraction. In effect, the 'speed of gravity' was infinite.

One thing attracted another and that was their inherent nature, without the need for any connection between the two. But by the nineteenth century, there was increasing suspicion that gravity worked by something moving through space – whether a flow of particles (as Newton had thought) or due to vortices in some kind of universal fluid – and if either of these were the case, gravity was likely to have a finite velocity. If gravity took time to get from A to B, it also opened up the possibility of having some kind of cyclical variation caused by gravity* – gravitational waves. But this idea would not be fully developed until Einstein had moved thinking on with his general theory of relativity.

Warps and wefts

The whole concept of spacetime – the mashup of space and time – is a little difficult to get your head around, but it was an essential requirement for Einstein to come up with his theory explaining gravity. In 1905, he had developed the special theory of relativity – special, not because it's wonderful (though it is), but because it dealt only with special cases where there was no gravity involved. Derived simply from Newton's laws of motion and the discovery that light always travels at the same speed, no matter how the observer moves with respect to it, the special theory showed that it was impossible to separate space and time. They are not distinct entities. For example, whether or not two events are simultaneous depends on how the observer deciding on

---

* If something has infinite velocity, there is no time for cyclic variation to take place, as it gets from A to B in no time.

their simultaneity moves with respect to them.* It's a bit difficult to visualise spacetime entirely as it requires four dimensions – the familiar three of space plus one of time. But we can think of ripples passing through a three-dimensional medium as a model for gravitational waves, which are ripples within spacetime.

Einstein had finalised his general theory of relativity in 1915, extending the special theory to include acceleration, which he had realised was impossible to separate from the nature of gravity. It was just the following year, 1916, that Einstein deduced that ripples in spacetime could in principle exist if massive bodies moved, making variations in their gravitational impact on spacetime. As a massive object moves, it drags the spacetime around with it, so the right kinds of regular motion would result a repetitive stretching and squeezing of spacetime. However, Einstein was also of the opinion that such waves would never be detected because the effects produced would be so incredibly small. Gravity, as we have seen, is a phenomenally weak force.

If gravitational waves existed as Einstein predicted, and they were to be detected, the device observing them would have to cope with variations that were a billion billion times weaker than the Earth's surface gravity – and this would be assuming some extremely powerful source of these waves. In practice, as we shall see, a number of possibilities exist for sources of gravitational waves that could reach this kind of level, but many physicists agreed with Einstein that these variations would be so insignificant – smaller than the

---

* Einstein illustrated this with an observer on a moving train, seeing two distant lightning flashes that are simultaneous for a stationary observer. The observer on the train will see the flash she is moving towards first.

gravitational effect of a passing car, for example – that they would never be detected. Einstein was so sure of this that he explicitly said that the waves were a theoretical possibility but would never be discovered.

Possible sources

Assuming for a moment that Einstein was wrong (he was occasionally) and it is possible to detect gravitational waves, only the most powerful sources are likely to produce anything that could be picked up. Likely generators of relatively powerful waves include the ripples produced by vast stellar explosions and the vibrations in spacetime of colliding black holes. In principle, the simplest form of gravitational wave should come from orbiting bodies. Many stars exist in 'binary' pairs which orbit each other – over 100,000 pairs have already been catalogued – and this motion should produce a continuous vibration, echoing out as a wave through spacetime. Even the Earth orbiting the Sun should have its own output – but this type of low-amplitude wave is likely to be far too faint ever to be detected.

The most likely source to be detectable – and that thought to have caused the signal detected in September 2015 – is a pair of 'inspiral black holes'. These are a pair of black holes (more on the nature of such pairs in Chapter 8) orbiting each other, where the orbits are decaying, bringing the black holes closer and closer together. In the last moments, as the black holes touch and merge, the massive bodies, moving quicker and quicker, should produce a rapid burst of powerful gravitational waves. These are sometimes described as a 'chirp', as the waves should shoot up in frequency and intensity as the bodies get faster and faster and finally merge, followed by a high-frequency 'ringdown' that rapidly fades away as the merged holes vibrate into a single body.

The other possibility for a signal strong enough to be detected is some kind of massive explosive disturbance in the cosmos, which sends a single burst of gravitational energy across the universe. Though not as well defined as an inspiral source, such a 'burst' of gravitational waves would consist of a very sudden, high-intensity gravitational shockwave that rapidly decays. The most likely source is a supernova, when a star explodes, particularly if it produces a gamma ray burst, but less work has been done as yet to model this kind of source, in part because they are expected to be much less frequent than inspiral events.

Gamma ray bursts are the most powerful individual events in the universe. They have been detected in distant galaxies (hence far back in history, given the time that light takes to reach us), and consist of an intensely bright flash of ultra-high-energy light (gamma rays), lasting between a fraction of a second and a few hours. It has been speculated that the bursts are associated with a supernova at the point of stellar collapse to form a neutron star (see page 57) or black hole, while some may be caused by energy released when neutron stars merge, again most likely forming a black hole.

A final source that has the potential of being detected is the gravitational wave equivalent of white noise, in the form of the so-called stochastic gravitational waves, a varied background hiss of subtle vibrations in spacetime that is thought may be traced back to the Big Bang. We already have many observations of the cosmic microwave background, which is light that emerged when the universe became transparent when it was around 380,000 years old, but in principle there could be background gravitational wave activity from all the way back to the Big Bang itself, as nothing stops gravity and so the universe has always been 'transparent' to gravitational waves. As we will see later, these background waves could have had a subtle effect on the polarisation of the early light and so be indirectly detectable in the microwave background pattern.

Tuning in

In principle, anything massive that moves through space generates gravitational waves, but two aspects of the waves – their amplitude and frequency – influence our ability to detect them. Amplitude is the 'height' of the waves – reflecting the amount of energy with which spacetime is vibrated. The bigger the masses involved and the stronger their interaction, the bigger that amplitude will be. This becomes particularly important when we look out billions of light years into space, far beyond our galaxy, as the amplitude drops off with the inverse of the distance. Yet we need to look out to cosmic distances, as events that will generate big enough amplitudes to be spotted are relatively rare. The more volume of space we can cover in looking for a source, the better – so one of the big differences in the new version of LIGO that spotted an event in 2015 was that the sphere of space it could detect events from was 30 times bigger than its predecessor.

Even small bodies generate waves – the partnership of the Earth and Moon for example, orbiting their centre of mass. (As it happens this centre is within the Earth, because the planet is a lot more massive than the Moon, so we tend to say that the Moon orbits the Earth, but in any dual system neither partner is stationary.) This means that the Earth–Moon system is losing energy via the gravitational waves it emits, and if this were the only influence on the Moon's orbit, the Moon would be spiralling into the Earth. In practice, though, the Earth has a much stronger tidal pull than its satellite – around 80 times as strong as the Moon's tidal forces that influence the sea (and the land) on the Earth. This imbalance slows down the Moon in its orbit. And this also influences the Moon's orbital distance, as a satellite can only travel at one speed in a particular orbit.

This happens because of a neat observation, pointed out to Isaac Newton by Robert Hooke, that being in orbit involves two separate activities. One is that the orbiting body falls towards the other body (just like Newton's semi-mythical apple falls to Earth). The other is that the orbiting body travels 'sideways', at a tangent to the other body, so that it always misses. This is the reason that astronauts float around in the International Space Station (ISS). They aren't in zero gravity – the Earth's gravity is still around 90 per cent of the surface level 350 kilometres up, where the ISS orbits. But the ISS is in free fall towards the Earth, meaning the astronauts don't feel the pull of gravity. It's just the sideways motion of the station that prevents it from crashing.

What this all means is that at any particular distance from the Earth, there is just one velocity an orbiting body can travel at that exactly balances out the tendency to plummet – and that's the speed you have to orbit at unless you are under power. So if the orbiting body slows down, as the Moon does due to the Earth's tidal force, it needs to orbit further away. Our satellite is gradually moving away, far faster than gravitational waves are losing its energy and causing it to get closer.

A system as small (in astronomical terms) as the Earth and Moon generates gravitational waves that are far too weak ever to be detected, which is why the LIGO observatories are looking for more massive systems, such as pairs of neutron stars or black holes, particularly just as they hit each other, where the gravitational distortions in spacetime will be the greatest.

Each type of event also generates a particular frequency, depending on the kind of action it is undergoing. LIGO can detect frequencies in the 30 to 6,000 Hertz range (these happen to be similar frequencies to audible sound waves, which is why gravitational waves are sometimes given that incorrect, if poetic description of interstellar sound). But to detect much larger black holes, for example, would mean moving to a lower frequency spectrum, below LIGO's receiving range. It is to access these frequencies that future observatories such as LISA (see Chapter 9) have been proposed, extending the frequency range that can be studied just as different kinds of electromagnetic telescope (radio, infrared, optical, ultraviolet, X-ray and gamma ray) make use of different frequency ranges of light.

Where are they?

Since Einstein's prediction of gravitational waves in 1916, then, we have been aware of the potential for them to exist. And despite all the problems of interference from local sources of vibration and the incredible sensitivity that would be required to pin down gravitational waves, physicists have been working on their detection since the 1960s. This is about pushing the limits of our technical capabilities and, as we will see, one physicist found that the challenge would have a devastating effect on his career.

THE GRAVITATIONAL WAVE CHALLENGE

4

The would-be inventor of a gravitational wave observatory faces immense difficulties because the waves are only likely to produce an incredibly small movement in an instrument, which will have to be detected. California Institute of Technology (Caltech) physicist Kip Thorne, who has worked on the theory of gravitational waves for many years and was one of the founders of the LIGO project, suggests that to understand how small these movements are, we start by looking at a familiar scale of distance with a centimetre measurement on a ruler. In his example, he uses the scientific standard for writing very small numbers where, for example, 10^{-4} indicates 1/10,000 – the '–4' superscript indicates 'divided by 1 followed by 4 zeroes'.

Thorne starts by dividing that centimetre by:

- 100 and you get the thickness of a human hair (10^{-4} metres [m]). Divide the distance by
- 100 again and you get around the wavelength of visible light (10^{-6} m). Now speed up and divide it by

- 10,000 and you get the diameter of an atom (10^{-10} m). Divide this by
- 100,000 and you get the diameter of the nucleus of an atom (10^{-15} m). Finally, divide that scale by
- 100 again and you get the magnitude of the *largest* motion (10^{-17} m) we might expect to see in the separation between the mirrors of a gravitational wave detector with arms a few kilometres long.

However, this isn't the only problem to face those who have been working on such devices for over 50 years. Even if you could detect a wave, there is also the difficulty of pinning down exactly where a source is in the sky.

Where did that come from?

Traditional observatories have a clear direction of action to work on. The receiving end of the telescope is pointed at a particular part of the sky, and that's where the observations come from. Simple. With gravitational waves, though, things are more complicated. Unlike the light-based telescope, there is no distinction to be made between up and down. Because nothing stops gravity, the waves could just as easily be coming up through the Earth as down from the sky.

One way to be sure of the direction of origin is to look for confirming activity with traditional telescopes. Once a putative gravitational detection has been made, conventional astronomers can be asked to point their telescopes in the direction the waves appear to come from, looking for supporting evidence in the visual, or radio, or other electromagnetic bands. However, this isn't possible without being able to have some idea of the orientation of the waves. And that is only possible with multiple detectors.

If the detector locations are at a good distance from each other, they will usually be at slightly different distances to the source – the delay between receiving a signal at one observatory and it reaching a second observatory can help pinpoint the direction the waves are coming from – and the more well-separated detectors that are involved, the more accurately it is possible to locate the source.

This is easier to envisage when considering a similar problem with sound. Imagine you are in a thunderstorm, but can't see any lightning flashes, so you are purely reliant on picking up the cracks of thunder as the electrical discharges blast through the air. We can roughly judge direction by using our two ears, because the shape of our ears means that sounds from the front are different from sounds from the back. But the gravitational wave detector doesn't have this option.

However, imagine we had a pair of microphones we could position well apart. Sound travels at around 340 metres per second at sea level. Let's say we place our microphones about a kilometre apart. Then, except in the unlikely situation that the lightning was exactly evenly spaced between the microphones, the sound would arrive at one microphone before the other. If we had placed the microphones roughly in line with the direction of the sound, we would get nearly a three-second lag between the signals. And from the difference in timing we would be able to tell if the sound was coming from behind or in front of us. The same applies to gravitational waves. However, the idea of having at least two detectors (and preferably more) has far more benefit than simply deciding the wave's direction of origin. It can also help deal with the false positives that inevitably arise in such a delicate system.

Trains, planes and automobiles

When the Jodrell Bank radio observatory was set up in the Cheshire countryside over 60 years ago, particularly after the construction of the great 250-foot steerable Mark I Telescope (now known as the Lovell Telescope), locals got used to engineers from the laboratory descending on their houses with the complaint that their washing machine or their vacuum cleaner were causing the scientists problems. Back then, electrical devices tended not to be as well screened against producing radio signals are they are today – and even now, some domestic appliances can emit noise in the radio band. When there was strong local interference from a radio source, it could easily drown out the intended observation, or even be confused for an actual signal.

It might seem very amateurish if a radio astronomer could not tell the difference between a pulsar (see page 57) and a dodgy spin cycle, but we have to bear in mind that radio astronomy observations are very different from those made with optical telescopes. Optical astronomers can look at a direct image of their field of view. Radio astronomers merely receive collections of numbers. These can be assembled by computer to construct an equivalent of the images from an optical telescope, but when dealing with nothing more than numbers, it is potentially far easier to confuse interference with a real signal than is the case with an image. And even in optical astronomy, it has not been unheard of to confuse an earthly light in the sky with a cosmic event.

Gravitational wave astronomers are in a similar quandary. Like the radio astronomers, all they have to go on are strings of numerical values, which can be interpreted as vibrations, usually represented as a side-to-side trace, despite the actual wave having a more complex three-dimensional form. These numbers are produced from small changes in the length of a detector, as we will explore in more detail soon. But these changes could also be brought about by local vibrations – anything from earthquakes to a plane flying overhead could produce a trace. As we have seen, even the gravitational effects of a nearby truck could be detected, these devices are so sensitive.

Each of the LIGO observatories in Louisiana and Washington state has a total of 100,000 channels of data that are being collected during an observing run. This information flow includes both the direct observational data from the interference of the laser beams in the tubes and detailed environmental data on the state of the mirrors, the content of the vacuum tubes and the environment around them. Seismometers feed into the mix, ensuring that any physical vibration of the Earth can be neutralised. This allows the scientists to maximise the chance of eliminating unwanted noise and interference, concentrating on any actual signals.

To be sure that local effects are ruled out, even more so than for directional considerations, there need to be at least two detectors in use, far enough apart that a vibration or moving mass local to one detector will not influence the other. With such a setup, the biggest remaining risk for confusion with an Earth-based disturbance is if, say, an earthquake happens roughly equidistant between the detectors, arriving at each site at approximately the same time.

Despite this apparently obvious issue, however, the first devices used in an attempt to detect gravitational waves were standalone – and before long, against everybody's expectations, a breakthrough detection of gravitational waves was reported.

Drinking in Weber's bar

It's hard not to admire the doggedness of the American physicist Joseph Weber. When he set up his initial 2-metre-long gravitational wave detector at the University of Maryland in the 1960s, it was hoped that it would be sensitive to 1 part in 100 trillion – an apparently tiny spatial shift, but it was estimated at the time that to detect a gravitational wave would require 10 million times more sensitivity. Yet despite being apparently so far out, Weber went ahead anyway. After all, the required sensitivity could only ever be an estimate – and even if it were correct, scientists can often learn a lot from the failure of an experiment (though they are just as upset when it happens as anyone else).

Weber's device was a 2-tonne cylinder of aluminium, about a metre in diameter to its 2-metre length, which was suspended from a sophisticated rig, intended to remove the possibility of an impact from vibrations producing fake signals. As gravitational waves passed through space they should also cause the material of the bar to expand and contract, which would then be detected by Weber's equipment. The hope was that the bar would amplify the signal by resonating with the frequency of the gravitational waves, rather in the same the way that a piano's strings can be made to vibrate of their own accord when sound of the correct frequency is played near to them.

Quartz crystals were stuck to the bar – as the bar expanded and contracted it would rhythmically squeeze and stretch the crystals. In materials like quartz, pressure has the effect of freeing up electrons from the crystal structure, starting a small electrical current flowing – known as the piezoelectric effect. By monitoring the currents produced by each of the crystals, the bar's movements could be tracked. To hardly anyone's surprise, nothing was detected. But Joseph Weber was not put off and redesigned his equipment, pushing up its sensitivity by as much as 100 times – though this still left it a factor of at least 100,000 too low to find anything on the sensitivity that was expected.

This was a very different and far more modest technology than the vast detectors that would be used in 2015. A 2-tonne cylinder may sound impressive – but this was a device that could fit in an ordinary physics lab, requiring a manageable budget that any university could stretch to providing. Weber was able to tinker with his equipment, producing a whole string of gravitational wave detectors, improving the electronics and trying different ways of protecting them from outside vibrations, with relatively few concerns being expressed about his modest financial requirements.

At the same time as improving sensitivity, Weber recognised the need to have more than one detector – as discussed earlier, this would both reduce the chances of a false reading and give some first suggestions of a direction for the source. Weber set up two bar experiments, one sited in Maryland and the other in Chicago about 1,000 kilometres (600 miles) distant. The physics world did not hold its collective breath – it still seemed highly unlikely that anything would come of these experiments. Yet in 1969, Weber was able to report a triumphant success and enjoyed a brief burst of fame. He had picked up a signal almost simultaneously on both bars, suggesting that it originated from an external source rather than a local vibration. Impressive though this was, however, there was one other possibility that Weber had to consider.

Imagine that gravitational waves didn't exist at all. Every now and then, each bar would produce a misleading signal. It might be due to a physical vibration, or even a glitch in the electronics, which by modern standards were crude, hand-soldered devices. And though it would not happen very often, occasionally these misleading blips would happen at roughly the same time on both detectors – even if each signal came from a totally different source. How would it be possible to eliminate such false positive detections? Weber devised a statistical technique to deal with them, effectively a method for deducing how likely a purely random coincidence would be, which could then give a level for the confidence that an apparent observation was significant.

Is this the real thing?

Note that by taking this approach, Weber had moved the observation from being something that actually *happened* – as, for example, when we see an eclipse taking place – to a statistical observation, something that had a good probability of not being a random occurrence but couldn't be definitively said to have happened. This has become a common way to determine the success of complex physics experiments, such as the detection of the Higgs boson, but it was far less familiar in the 1960s, and was an approach that astronomers had traditionally not needed in their armoury.

The way that Weber's technique for testing random coincidences worked – a methodology still used in a modified form by today's gravitational wave astronomers – was to shift the readings from one of the detectors in time. Imagine each reading was represented by a long roll of paper, with the vibrations it detected marked on it, like the paper roll from an old-fashioned seismometer. Then as one sheet of paper was slid along next to the other, a pair of signals, where a strong vibration stands out from the background noise, would eventually line up. This pair would definitely not be a true detection, as the two signals occurred a considerable distance in time apart. Their apparent match from the 'time slide', as the process became known, was a purely random event. And by continuing this process, making the slide over and over again, Weber could work out how frequently an event on both detectors was likely to coincide in a particular chunk of time – hence getting a feel for the likelihood that a detection was spurious.

Combining this statistical analysis with the evidence he believed he had collected of simultaneous signals on the two detectors, Weber deduced that his detection was, indeed, real. Against all the odds, he had found a gravitational wave. This was exciting, but hardly definitive. It's a bit like the kind of study that is always being reported in the newspapers telling us that, say, drinking red wine is good for us (or bad for us). These articles tend to be based on a single study. But it's rare that a piece of work like this from a single lab with a limited number of observations is sufficient to be sure of the outcome. Scientists believe that it's important to duplicate an experiment elsewhere to ensure that they get the same results. And the combination of a cheap and cheerful piece of technology with the potential prize of a whole new form of astronomy meant that it wasn't long before a whole string of universities were working on their own Weber bars in an attempt to duplicate Joseph Weber's results. The trouble was, though, that no one managed to reproduce the outcome. No one else got a detection.

At the same time, the theoreticians were working on models of possible sources. Gravitational wave detection is an interplay between the work of the detector-building experimental physicists and the pencil-wielding theoreticians (though these days theoretical physicists are more likely to be wrangling complex computer programs). The experimenters may be able to build equipment that can respond to intensely subtle spacetime vibrations, but of themselves, the results that those detectors produce tell us nothing about the universe. Weber didn't really care about this – all he was interested in was the detection process. He was happy to leave an exploration of how the waves might be generated to others. But when the theoretical physicists got to work on his results, the numbers didn't seem to add up.

More accurately, they added up to far too much. Any such detector has a threshold, below which it won't detect a vibration. Many had assumed that Weber's bars would only detect gravitational waves that were more powerful than any that were predicted to exist. When the theoreticians started to search for a potential source for waves of this power out in the universe, they were stumped. Of course, it was entirely possible that the theoretical physicists had missed something. But, more worryingly, British astrophysicist Martin Rees, later the Astronomer Royal, and his colleagues showed that if gravitational waves existed on the scale suggested by Weber's findings, they would require a source with so much energy that it could blast the entire Milky Way apart.

Not only were Weber's results impossible to reproduce, according to Rees they didn't even make sense according to theory. Admittedly, this view was not shared by all theoreticians. The US-based British theoretician Freeman Dyson pointed out that some potential sources could produce a specific frequency that might generate waves that could get Weber's bars to vibrate with a lot less energy because the bars were particularly sensitive to that frequency. Specifically, he came up with the concept of two very compact stars, orbiting each other closely, as a potential source. Although Weber's detections seem not to have been genuine, this idea of Dyson's would turn out to be by far the easiest detectable source of gravitational waves for later, more sensitive detectors.

If there had only been one or two other experiments, also managing a detection, Weber's results might have held up better. But by the early 1970s, Weber bars were set up in many universities in the US and Europe – to no avail. Weber, once the golden boy of gravitational waves, began to be regarded with suspicion. There was even a Moon-based experiment derived from Weber's work, set up by NASA in 1972 as part of the Apollo 17 mission. The lunar lander carried a pair of gravimeters – gravitational strength detectors. One of these, designed to discover more about the Moon's geological structure, proved effective, but the 'lunar surface gravimeter', specifically designed to follow up Weber's experiments, failed to deliver. Inevitably, with the mass and size limitations of an experiment carried by an

Apollo lunar lander, this couldn't be a conventional 2-tonne Weber bar and instead was a kind of spring balance with a small detecting mass suspended from it. It was hoped that passing gravitational waves would cause the mass to make slight movements, disturbing the balance, but nothing was observed.

We shouldn't minimise Weber's contribution. He really did kick-start the search for gravitational waves at a time when no one else was taking the possibility of finding them seriously, and he devised the technique to check for coincidences in detections that it is still used today. Although his 'successful' observation was almost certainly a mistake, it made it more acceptable for others to invest time and energy in searching for the phenomenon.

Admittedly, Weber made some serious errors along the way, justifying gravitational wave scientists' doubts about his results. At one point, he claimed to be getting signals arriving 24 hours apart, which he put down to the rotation of the Earth – but it was pointed out to him that as gravitational waves would pass through the Earth as if it is not there, they should have come at twelve-hour intervals. Worryingly, when he was told this, he somehow managed to rework the data to show the correct twelve-hour gaps. On another occasion, he claimed a successful dual detection with another lab's bar, failing to notice both that the other device was in a different time zone – so the detections weren't actually simultaneous – and that the waveform detected at the other site was a test signal, generated by the lab, not a real source. However, these setbacks did not stop Weber from continuing to work on gravitational detection to the end of his career. And he next proposed to make a detector that was more sensitive by making it supercool.

Chilling the bars

We've focused at some length on the possibilities of an out-side disturbance setting up a vibration in a bar. Whether it's passing traffic, earth tremors or the thunder of students' feet on corridor floors, it was essential that the Weber bars were isolated from their surroundings. This wasn't helped by the relatively small and cheap nature of the equipment, meaning that the bars tended to be set up on university campuses rather than at isolated, low-interference sites as is usually the case with modern observatories. But no matter how much the bars were cushioned against noise from external vibra-tion, they could still do their own shaking.

Motion is the natural state of matter. Nothing in the universe is completely still for any length of time. The size-able, heavy metal bars that formed the detectors might have appeared to be entirely static, but zoom in to view them on the atomic scale and the atoms that made them up would be seen to be jiggling about rapidly with the thermal energy that gives them a temperature. We might think of temperature as being just something we measure with a thermometer, but fundamentally temperature is a measure of the kinetic and potential energy of the atoms that make up a substance. The more the atoms jiggle about, the higher the temperature is. This is why, if the temperature gets high enough, those locked-in atoms can break away from the electromagnetic bonds that hold them in place and the solid melts.

However, what physicists wanted to do to avoid noise in their bars was not to increase the amount of movement in the atoms, but to decrease it. Although the thermal energy of the atoms is randomly directed, so overall it will typ-ically not result in any motion of the entire bar, the way it combines is statistical. Occasionally, in part of the bar, more atoms will pull in one direction than another and the result with be a ripple passing through that section of the material – a tiny vibration that originates from within the bar. You would never see this simply by looking at the solid object. But at the level of sensitivity employed by Weber's equipment and that of his contemporaries – by now it was capable of detecting a change in length around the size of the nucleus of an atom – such infinitesimal changes might show up as a spurious gravitational wave detection. Hence the need to be supercool.

Just as raising the temperature results in an increase in the amount of jiggling of the atoms, a decrease in temperature reduces that motion. In principle, if an object could be taken all the way down to absolute zero, which is –273.15°C, its atoms should be totally devoid of kinetic energy. In practice, this isn't possible to achieve, as the uncertainty principle of quantum physics tells us that it's impossible to exactly know both the location of a quantum particle and its momentum simultaneously – something that could be deduced if a particle were totally static. But it is possible to cool an object down to within a fraction of a degree of absolute zero. At the time of writing, the lowest temperature ever achieved is around 0.004 kelvin or –273.146°C.

Although the scientists working on Weber bars at the start of the 1970s couldn't achieve such near-perfection, they hoped that by reducing the temperature to just a few degrees above absolute zero, they would be able to eliminate any significant thermal vibration and reproduce Weber's results with less chance of a false positive. Weber himself did not complete a working low-temperature bar before his death and, unfortunately, although some were built, they did not deliver. Never again would Weber bars produce what appeared to be a real observation.

A totally different type of detector would need to be deployed if there was to be any hope of making gravitational wave astronomy a reality. But long before that was possible, two US scientists would make an observation that confirmed the existence of gravitational waves. Remarkably, this feat, for which the pair would win the Nobel Prize, was performed without ever detecting the waves.

DANCE OF THE NEUTRON STARS

5

In 1974, American radio astronomers Russell Hulse and Joseph Taylor were running a series of observations from the vast, fixed-dish radio telescope located at Arecibo in Puerto Rico when they made a striking discovery. The regular variation in the 'tick' of a celestial clock was speeding up. And that just shouldn't happen.

The detection in the jungle

The Arecibo radio telescope is a monster of an instrument. At the time of writing, the largest steerable radio telescope is the Green Bank Telescope in West Virginia, USA, which has a 100-metre-diameter dish. At the UK's Jodrell Bank observatory, the Lovell Telescope, which celebrated its 60th anniversary of powered operation in 2017, is now the world's third-largest steerable device at 76.2 metres in diameter. By contrast, the Arecibo telescope is 305 metres across. For a long time this made it the world's largest telescope, though China's

Five-hundred-metre Aperture Telescope, known as the 'eye of heaven', beat it by a significant margin in 2016. (Strictly, Russia has an even larger radio telescope, RATAN-600 with a radius of 576 metres, but this only has a thin section of reflectors around the rim, rather than a complete dish.)

The location of the Arecibo telescope is striking, forming a circular clearing in a Puerto Rican forest. The original construction was helped by a natural sinkhole in the karst, a geological formation where more soluble rocks have washed away over time to form a deep depression. Because of its sheer scale, the dish of the Arecibo telescope cannot be moved around to point to different parts of the sky. It is made up of thousands of aluminium panels, laid out on steel cables suspended above the ground. However, this doesn't mean that Arecibo is limited to receiving radio waves from a single direction. Suspended 150 metres above the dish is a platform where the receiver is based, collecting radio waves that are reflected by the dish. A mechanism allows the platform to move receiver aerials around above the dish to capture signals across 40 degrees of the sky.

This massive device was originally intended for research into the Earth's ionosphere, but has been deployed for a number of projects, including SETI (the Search for Extraterrestrial Intelligence), where it was used to search for alien radio signals. As Arecibo has transmitters as well as radio receivers, it has also been used to send out a focused signal into space towards the M13 cluster, in the hope of reaching other civilisations. However, Arecibo's most significant discovery in its 50+ year history has been Hulse and Taylor's work.

In 1974, the astronomers found a strange pulsar 21,000 light years from Earth, which was given the uninspiring name PSR1913+16. And what was initially considered special – in fact, downright odd – about this pulsar was that its pulses speeded up and slowed down. That wasn't supposed to happen.

Lighthouses in space

Pulsars are an astronomical phenomenon of the radio telescope era. The first pulsar, discovered at Cambridge in 1968 by PhD student Jocelyn Bell (now Bell Burnell) and her supervisor Anthony Hewish, was given the nickname LGM-1. The tongue-in-cheek designator stood for 'little green men', then a popular term for aliens, so strangely mechanical was the signal that this object produced. It gave off regular pulses of radio waves, like the ticking of a distant cosmic clock – or a deliberately produced, automated beacon. It was later realised that what was being observed was a neutron star (more on these in a moment) rotating at high speed, its radio output sweeping around like the beam of a lighthouse.

Many more pulsars have since been discovered, with frequencies that range from seconds down to just a few thousandths of a second. Bearing in mind this means that these stars have a 'day' that is just milliseconds long, they have to be rotating at an immense speed. At first that speed seemed impossible. It was only when it was realised what pulsars had to be made of that such a speed of rotation seemed even vaguely possible. For the fastest known pulsars to date, the equator of the star would have to be rotating at around a quarter of the speed of light.

The possible existence of neutron stars had been predicted back in the 1930s, when the neutron particle was first discovered in atomic nuclei. But Jocelyn Bell's discovery of a ticking body in the sky was the first evidence that these bizarre stellar structures really existed. When a large star, at least eight to ten times the mass of the Sun, nears the end of its life, it has the potential to become a neutron star. As the star runs out of fuel for the fusion reactions that power it, it starts to collapse under the force of gravity. Up until then, the energy of those reactions had puffed the stellar material up, but now there is little to resist gravity, which has a powerful effect with so much mass involved. Because they are so far away, we don't really grasp how big stars are. It's worth remembering that the Sun – a lot smaller than a potential neutron star – contains 99.9 per cent of the matter in our solar system.

With the supporting energy dispersed, the old star begins to collapse in on itself. This 360-degree avalanche of stellar matter generates a huge surge of heat, which blasts off the star's outer layer, making it a supernova that may well briefly flare out with the brightness of a whole galaxy. But what is left of the star after the outer layer blows off, a remnant typically between 1.1 and 2 times the mass of the Sun, is a ball of neutrons that continues to collapse until it is only 10 to 20 kilometres across. At this stage, a quantum effect known as the Pauli exclusion principle prevents further collapse. But the resultant neutron star, that 10- to 20-kilometre ball, is a remarkable beast.

Ordinary matter is mostly empty space. Each atom consists of a tiny nucleus with a cloud of electrons well spaced out from that central core. The nucleus, consisting of relatively heavy protons and neutrons, is like a pea in the volume of a cathedral-sized building, with the rest of the atom empty space, apart from the cloud of lightweight electrons in the outer reaches. But a neutron star consists only of neutrons.

With no electrical charge to repel each other, these particles can be pulled closer and closer by gravity until the exclusion principle kicks in when they're practically on top of each other, enabling that great mass to be squeezed into a ridiculously small space. The result is that a teaspoonful of neutron star material would weigh about 100 million tonnes. And the other effect of this extensive compression process is the remarkable speed of rotation that is observed in pulsars.

The starting point for this frenetic whirling is that pretty well everything in the universe rotates to some degree. This is an inevitable consequence of the way that stars, solar systems and galaxies are formed. If you imagine a cloud of dust and gas being gradually pulled together by gravity to form a star, unless that cloud is perfectly symmetrical – which it never will be in practice – one side will have slightly more matter to provide gravitational pull than another, meaning that the dust and gas will spiral as it moves in, rather than flowing radially in straight lines. This produces an initial rotation in a star. And if that star collapses to form a neutron star, the rotation is magnified by the ice skater effect.

If you've ever seen an ice skater spin, they start with their arms out, then suddenly pull the arms in, at which point they begin to rotate much faster. You can try this out yourself if your local playground has the apparatus where you stand on a small platform that rotates around a vertical bar. If you start the platform spinning with your body pushed out at arms' length from the platform, then pull yourself quickly in to be upright against the bar, your rate of spin will accelerate dramatically. It's a matter of conservation of angular momentum.

Like energy, and a number of other properties, the amount of angular momentum in a system stays the same. This is the amount of vigour with which something rotates. And as angular momentum increases with both the speed of rotation and the distance of the mass in the object from the centre, if you move the mass inwards, reducing that distance, the speed of rotation has to go up. Bearing in mind that a neutron star will have collapsed from being perhaps 1,500,000 kilometres across to, say, just 15 kilometres, this effect has an intense impact on the speed of rotation of the star.

Stars usually have a strong magnetic field and, as the neutron star collapses, this will become intensified at the surface of the star. As particles in space are pulled in by the powerful gravitational force, they will be accelerated and give off radiation which is funnelled by that magnetic field. This means that a neutron star can have a powerful beam of electromagnetic radiation emerging from its poles, like great jets of light.

We're used on the Earth for the magnetic and rotational poles of the planet to be approximately aligned (though magnetic north is not quite at the North Pole), but the magnetic poles of neutron stars need not be lined up with the rotational poles. If there is a considerable angle between them, the beams from the magnetic poles will sweep through space at the speed of rotation of the neutron star – and if Earth happens to be in the direction of one of those beams, we will see the neutron star producing a 'flashing' pulse of radio waves as the beam repeatedly sweeps by.

From spin to gravity

Pulsars, then, are fast-spinning neutron stars giving off radio waves in lighthouse beams that appear at Earth as a series of high-speed blips in the radio spectrum. Their discovery rightly won the 1974 Nobel Prize in Physics for Anthony Hewish, though Bell, who actually found the first pulsar, was omitted from the award. This made Hewish's Cambridge colleague, astrophysicist Fred Hoyle, issue a furious protest, though Bell herself accepted being passed over with good grace. In 1993, there was no such controversy when Hulse and Taylor shared the Nobel in Physics for 'the discovery of a new type of pulsar, a discovery that has opened up new possibilities for the study of gravitation'. Specifically, they were able to deduce the presence of gravitational waves from their observation of a pulsar with variable rotation rates, even though they did not directly observe the waves themselves.

That may seem like a remarkable piece of sleight of hand, but indirect observation is something that we often do, both in physics and in the everyday world. If, for instance, we see a piece of paper moving 'of its own accord', we can sensibly deduce that there is a breeze that is moving it. We can't see the current of air that causes the paper to move, but we can observe the outcome of its existence and make a deduction. Of course, making that deduction is not as straightforward as a phenomenon we can directly experience. In principle, there could have been another reason for the piece of paper to move, such as an earthquake. But we can usually connect it to an air current. Similarly, Hulse and Taylor did not 'see' gravitational waves but they were able to deduce the presence of the waves from the impact that they appeared to have on not a single star but a pair of neutron stars, one of which was a pulsar.

What first attracted their attention in 1974 was that

PSR1913+16 was a pulsar that didn't have a constant 'tick' rate, but sped up and slowed down every few hours. It was hard to imagine how a normal pulsar could have rapidly varying rotation rates. The spinning neutron star is like an incredibly massive flywheel – which makes it very difficult to make significant changes to its rotation speed over short timescales. So when it was observed that the pulsar appeared to be doing exactly this, it seemed far more likely that a pair of stars was involved, orbiting around each other. One was, indeed, a pulsar, but the other was a non-radiating neutron star, and the orbit of the two stars was causing the observed pulse rate to vary as a result of the Doppler effect, the same effect that causes the sound of a moving siren to change in pitch as it passes us by.

Imagine what would happen when the pulsar was in the part of its orbit where it was moving towards us. First it would send out a pulse. By the time it pulsed again, it would be a bit closer to the Earth – and so the radio waves for the second pulse would take slightly less time to get here than the first one. This pulse would arrive earlier than it would if the neutron star were stationary with respect to us. The effect would be that while the star is moving towards us, the pulse rate would be increased. When the pulsar was in the part of its orbit where it was moving away from us, the opposite would happen and the pulses would have longer gaps between them than we would expect for a stationary star. The result would be that on a regular basis, the pulsar would appear to speed up and slow down: from the observed pace of the change, it was thought that the pulsar and its companion neutron star in the formation were spinning around each other every eight hours.

Two massive bodies like neutron stars orbiting each

other provides exactly the sort of system that the general theory of relativity predicts should generate gravitational waves, though the relatively low mass of the neutron stars compared to black holes would mean that the waves would be too weak for any existing Earth-based system to detect them – particularly back in the 1970s when Hulse and Taylor were working. But while the waves themselves could not be detected, their production was predicted to have a side-effect. Generating such waves takes energy, which has to come out of the orbits of the neutron stars. The loss of energy from the binary neutron star system should have had an impact on the orbits. As the rotating pair of stars pumped out those waves, they should have fallen inwards towards each other, very slightly. And this would have made them orbit at a slightly faster rate. If the system were emitting gravitational waves, the orbital rate should be observed to speed up.

If that were the case, you would expect that the frequency *of the variation* of the neutron star's pulse – reflecting the star's journey around its orbit – would also increase over time because the time between the pulsar being in the part of the orbit where it is moving towards us and the part where it is moving away from us would reduce. And that is exactly what Hulse and Taylor observed happening in the pulses from PSR1913+16. The change was slight, but it was exactly what would be predicted by the general theory of relativity. It was not definitive proof of the existence of gravitational waves – the waves had not actually been detected, and it was always possible the decay of the orbits could have another cause – but the observation was very strong supporting evidence for the existence of such waves.

What is a direct detection?

When the LIGO team was trying to sort out a title for the paper describing their discovery of gravitational waves, 41 years after Hulse and Taylor's original discovery, they agonised over the use of a particular word that distinguished what they had done. Was it 'direct'? From the speeding up of the variation in pulsar frequency over time, the astronomers in 1974 had made the *indirect* assertion that gravity waves existed. But the LIGO discovery would be a *direct* observation, picking up the waves generated by the source themselves. And, arguably, as a result, their observatory would be making a direct observation of the body or bodies producing the waves. Of course the LIGO scientists couldn't *see* those bodies in a literal visual sense. However, just as an optical telescope picks up the light waves produced by a distant object, the LIGO observatory was picking up the gravitational waves from a distant object and so had every bit as much right to be called a direct observation of the source.

This direct aspect would be of particular interest because, as we have seen, the source of the first LIGO observation was thought to be a pair of black holes – cosmological entities that had never before been directly detected. Though we think we know a lot about black holes, it is all based on either theory or indirect observation of, for instance, the effect a black hole has on nearby visible objects and material. Until the gravitational wave event of 2015 we had never directly detected a black hole. Our relationship with some of the most remarkable inhabitants of the universe was about to change.

But before we can find out more about black holes and their nature, we need to have sensitive enough equipment to be capable of detecting those gravitational waves. Apart from Weber's early observations, now thought to have been made in error, Weber bars had never succeeded in making a detection. It was time to deploy a whole different technology. And though the implementation itself would end up stretching twenty-first century technology to its limits, the approach that would prove effective would turn out to be one that had played a part in Victorian science by showing that the medieval-sounding concept of the 'ether' was not needed.

The interferometer was about to take its rightful position centre stage.

MAGIC MIRRORS

6

As we saw in Chapter 2, the idea that light was a wave had been given a big boost in the nineteenth century because it was discovered that beams of light produced interference patterns, where, depending on the relative position, two waves of light can either reinforce each other and become stronger, or cancel each other out. This effect would prove to be an extremely precise way of measuring changes in distance – and hence a valuable tool in the hunt for gravitational waves.

Running interference

Imagine a simple piece of apparatus – conceptually simple at least, though practically challenging to build. We take a beam of laser light – light that has a single, precise wavelength – and split that beam into two. This requires a device imaginatively known as a beam splitter, which at its simplest is a part-silvered mirror which reflects some of the light and allows part of it through. (In optical experiments, beam splitters are usually more sophisticated, constructed from combinations of prisms, but the outcome is the same.) The light from the beam splitter is sent off in two directions at right angles to each other. At the end of its travel, each beam hits a mirror, which sends the beam of light back towards its point of origin to merge with its fellow, before being projected onto a screen where the resultant interference pattern can be observed through a microscope, or (more likely in a modern experiment) will be picked up by a light-sensitive detector. The result is something like this:

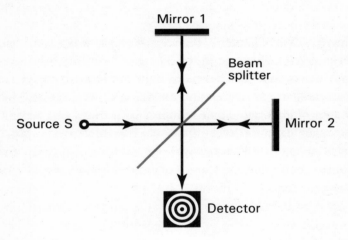

Because not even a laser has perfectly parallel rays, but spreads out a little by the time it reaches the mirror, some light waves will travel slightly further than others, so the result is an interference pattern when the beams are combined at a detector, with dark circles where the waves from the two mirrors cancel each other out and light circles where the waves from the two mirrors reinforce each other. But if either mirror moves closer or further away from the beam splitter, those fringes will shift, as the distance the light has to travel will have changed, and so it will be at a different position in its wavelength when it meets its counterpart before arriving back at the detector.

For a change in the interference pattern to be detectible, the change in the path length only has to be a fraction of the light's wavelength, and as a full wavelength is (for visible light) between 400 and 700 nanometres (billionths of a metre) in length, this means that shifts on a nanometre scale can be measured.

These devices, known as interferometers, have proved useful in measuring changes in distance very accurately, and until LIGO's remarkable success, they were best known for their role in changing our understanding of the way that light works. As we saw in Chapter 2, when light was first discovered to act as a wave it was assumed that there had to be a medium, even in empty space, for it to wave in, called the ether. Despite James Clerk Maxwell's discovery that light was an interplay between electricity and magnetism and hence had no specific need of an ether, many – including Maxwell – continued to think that this strange, diffuse but rigid substance existed. However, an experiment undertaken towards the end of the nineteenth century using an interferometer, which was meant to demonstrate the existence of the ether, instead made it highly likely that it didn't exist at all.

The missing ether

The interferometer experiment was undertaken by US physicists Albert Michelson and Edward Morley in 1887 at the

Case School of Applied Science (now Case Western Reserve University) in Cleveland, Ohio. Their equipment was similar to the diagram on page 68, except it had multiple mirrors on each path, so the beams travelled further, back and forth, before they were merged to cause interference. This device was mounted on top of a heavy slab of stone over a metre across, supported by a wooden frame that floated in a trough of mercury so that the whole thing would keep moving at a steady pace once it started to rotate. The setup was carefully designed to minimise friction: once the stone 'table' was travelling at around one rotation every six minutes, it would keep going for hours. Rather than look for a displacement in the mirrors, which were fixed in place, though, the experimenters were looking for the impact of the ether 'wind' on the time light took to cover the distance.

The idea was that as the Earth moved through the ether, light that was travelling in the same direction as the planet would have to fight against the motion of the invisible fluid in the opposite direction, while light at right angles to the movement through the ether would not be influenced. At any one time, one of the beams would be pointed more into the direction of the ether wind and the other across it. As the instrument turned, the interference fringes formed by the meeting light beams, studied through a microscope against a grid pattern, should shift, detecting the impact of motion through the ether. But try as they might, Michelson and Morley could not discover any effect. If the ether was really there, it was well concealed. And despite years of attempts to repeat the experiment by the Cleveland pair, all the way through to 1929, and further attempts by other scientists elsewhere, nothing would ever be detected.

Strictly, this research could not prove that the ether did not exist. As the old saying goes, absence of evidence is not evidence of absence. But when time after time such sensitive equipment showed no effect occurring, it was hard to argue that the ether was really there. In practice, there were attempts to explain the lack of detection. George FitzGerald and Hendrik Lorentz suggested that the arms would change in length due to their motion, negating the effect of the ether wind – in practice, this contraction could not save the ether, but it was one of the contributory factors to Einstein developing the special theory of relativity.

In many ways, the search for the indirect effects of the ether using an interferometer is a mirror image of the search for gravitational waves 100 years later. For many years, the gravitational wave researchers also failed to find anything. But there was a big difference. The effect of the ether wind, had it existed, should have been clear and easy to find – so, after many failed attempts, it was reasonable to dismiss the existence of the ether. There was no need for it. But even with the most sensitive interferometers, the impact of gravitational waves was expected to be minuscule and nearly impossible to detect. This meant that it seemed more reasonable – if still highly frustrating – to keep plodding on despite year after year of failure. Even so, the use of interferometers meant that, for the first time, it was at least conceivable that gravitational waves could be detected.

Interfering with gravity

When gravitational waves pass by, spacetime is stretched and squeezed alternately in two dimensions at right angles to the wave's direction of travel. If a gravitational wave were to pass through an interferometer, the result should be different on the two arms which are at right angles to each other, an effect that changes repeatedly as that two-dimensional stretching and squeezing occurs. When one arm gets slightly longer and the other slightly shorter, there will be a movement in the interference pattern because one light beam has to travel slightly further than the other. As the cause is an oscillating wave, the outcome should be a distinctive vibration in that interference pattern. Of course, being sure of exactly what is involved would not be trivial. As we have seen, the smallest physical vibration in the instrument could result in a false reading, while a wave that crossed both arms, by far the most likely occurrence, could have unpredictable results. Even so, interferometer-based observatories had the potential for far more sensitivity than a Weber bar, and so were far more likely to achieve a detection.

At the time of writing there are five main gravitational wave observatories worldwide. As well as the pair that form LIGO, there is the French/Italian VIRGO interferometer located at Cascina, near Pisa in Italy, nearly as large as LIGO's observatories with 3-kilometre arms, the German/British GEO600 with 600-metre arms, and the Japanese TAMA 300 at half that size. Although it might seem that the individual detectors are limited in use compared with LIGO's paired observatories, as it would be impossible to distinguish a local vibration from a detection, in practice all the gravitational wave observatories share data, so the combined devices can effectively act as a single world-spanning observatory, giving greater accuracy and more ability to pin down the direction a wave has come from.

These aren't the first interferometers to be constructed in

the hunt for gravitational waves. A pair of Russian scientists, Mikhail Gertsenshtein and Vladislav Pustovoit, had put forward the basic concept in the early 1960s, but discovered that the practicalities of making such a device useable defeated them. The principle was fully conceived back in 1972 when American physicist Rainer Weiss, working at MIT, wrote a report on a theoretical approach using technology similar to the Michelson Morley experiment, though fixed in place rather than rotating, and with much longer arms. Weiss had first come up with the concept when teaching a course on general relativity in the late 1960s, where he admits he was only one step ahead of the students in a subject he knew little about. He set his students a challenge to keep them interested, asking them to devise a gravity wave detector based on a Michelson interferometer.

On the edge of possibility

Weiss took a big step forward in making an interferometer detector practical by thinking through the implications of attempting to measure tiny variations in the lengths of laser beams as space and time were squeezed and stretched by a passing gravitational wave. To make the device practical, he had to identify the major noise sources – in effect, figuring out what the competing vibrations would be, so that an effective gravity wave detector could rule them out. Weiss devised ways to deal with such unwanted inputs, particularly by constructing sophisticated suspension mechanisms for the mirrors placed at the ends of the laser arms, and worked out, as a result, the level of sensitivity that such a device could achieve.

From his calculations, Weiss deduced that a working gravitational wave interferometer would need very long arms – kilometres in length, rather than something that could be run on the lab bench – so that it was able to pick up the tiny vibrations in space that were gravitational waves. Journalist Janna Levin quotes Weiss as saying: 'I didn't like big science. But I could only do [the experiment] if it was a big project. There was no other way to continue. The science required it. I didn't ever believe you could build it small.'

It's impossible to overstate just how delicate a measurement is involved in an experiment like LIGO. Kip Thorne tells of how he first came across the ideas from Weiss's report when he was co-authoring a book on gravity, simply titled *Gravitation*, with Charles Misner and Thorne's former PhD advisor, John Wheeler. In the book, Thorne practically dismissed the idea of successfully detecting gravitational waves.

The reason, Thorne points out, becomes obvious if you home in on the scale of movement in the mirrors that has to be detected. He said: 'I just looked at it and I looked at the numbers and it was obvious this was not a very good idea. We were about to go to press [with *Gravitation*] and I wrote in there that this is not a very promising approach.' Thorne simply did not believe it would be possible to detect movements smaller in scale than the nucleus of an atom.

Both Germany and the UK had small interferometers with arms varying between 3 and 30 metres working in the 1970s and 80s. All these devices had one thing in common. They never discovered anything and were never expected to do so. But they provided both testing grounds for the extremely advanced technology required to keep the mirrors stable for the detection of gravitational waves, and enabled the scientists working with them to eliminate the lower reaches of possible detection – in science, the absence of an observation can itself provide valuable data. Gravitational waves were enough of a theoretical concept that it was always possible that nature would provide a surprise.

The LIGO challenge

Kip Thorne and his colleagues worked through the details of constructing a pair of large-scale interferometer-based detectors, making the decision to go ahead between 1976 and 1978. This was no easy project to get off the ground. Leaving aside that phenomenally subtle measurement required – beyond anything that had been achieved at that date – and the likelihood that the first version of a large interferometer observatory would still not observe anything because it was unlikely to reach the desired sensitivity, there was the fallout from Weber's work. With his bars and his widely announced discovery, Joe Weber had briefly become a scientific celebrity – but when his results could not be reproduced and things began to go wrong for him, he had fallen so far out of favour that the whole concept of searching for gravitational waves seemed to be tainted by history.

A Caltech team under Kip Thorne worked on a 40-metre prototype interferometer to iron out practical issues, while Rai Weiss's team, based at MIT in Cambridge, Massachusetts, built a smaller test rig to iron out any mirror support issues and was also tasked with putting together a feasibility study for a full-scale implementation. In essence, this was a matter of looking for anything that could possibly go wrong with the project and suggesting solutions for these problems. This was not just about the physics, but required an in-depth knowledge of the practicalities of dealing with a project on this scale, from the steelwork in the arms of the interferometer to the costings needed before it was possible to put in an application for funding.

A good example of the tiny details that the feasibility study had to deal with was the fact that the steel pipes making up the arms would not be entirely inert (even without the addition of animal urine). To enable the laser beam to pass up and down the arms of the interferometer between the pairs of mirrors, the air had to be removed from the tubes. If there was any gas inside, the laser's beam would be scattered and there would be movements introduced by the random motion of the air molecules, constantly unsettling the system. But when the air was pumped out to evacuate the pipes, small amounts of hydrogen would be released from the steel into the beam, itself a potential cause of error. Weiss's team had both to anticipate the presence of the hydrogen and work out how to deal with it, allowing for the variation in readings that would be produced by the level of hydrogen that was released.

In 1983/4, the two groups from Caltech and MIT presented their findings and plans to the US National Science Foundation. Because of their careful planning they achieved a relatively easy funding decision at a time when the US government was beginning to have doubts about large-scale physics experiments. The estimate given at the time was that the two observatories would cost around $100 million, but this was only for the basic construction. No one expected a fully functional LIGO project to come in so cheap, and the reality would be over ten times as expensive. Initially the project was run by a triumvirate of Thorne, Weiss and another experimental physicist, Ron Drever, who was brought in from Glasgow University, a European centre of expertise in gravitational waves. Thorne has remarked that the two experimenters 'rarely saw eye to eye', and notes that the three of them were referred to as a troika, which 'at the time referred to a very dysfunctional leadership and we were as dysfunctional a leadership as you could imagine'.

A dysfunctional leadership

Certainly the three scientists were very different in temperament. Weiss was a practical, get-things-done guy. Thorne, the theoretician, was methodical and sometimes slow to make decisions. By contrast, the Scot, Ron Drever, was something of a whirlwind. Science writer Janna Levin says that 'Ron would release a deluge of ideas on his team each new day. Ideas were abundant, but decisions were scarce.' There was no doubt that Drever came up with some very important ideas in the development of gravitational wave astronomy, without which a functional LIGO might never have got past the planning stage – but Drever was not happy for any control to be imposed on his contribution. His approach was fine for the initial backroom activity of idea generation, where possibilities can be continuously flung out and debated, but it was not ideal as a model for the working of the management team of a major construction project.

It was Drever, for example, who brought in a different approach from the Michelson Morley experiment that would powerfully increase the interferometer's chances of making a successful detection. In the original Michelson Morley setup, a series of mirrors allowed the beam to pass back and forth along each arm four times before being combined. The Weiss design for a gravitational wave detector used suspended mirrors at the end of each arm, with the beams oriented so they would bounce back and forth between the mirrors dozens of times at slightly different points on each mirror every pass, before being collected and merged. This back and forth repetition effectively amplifies the effect of any small movement of the mirror. But Ron Drever proposed taking things far further.

It was Drever who suggested that they make each arm of the interferometer into a vast Fabry-Perot cavity. In effect, these devices are like enlarged versions of the cavity in a traditional laser, where light passes into the cavity and then bounces back and forth in exactly the same direction, each pass overlapping with the next, building up in power as it goes until the beam is released. The cavity acts as a kind of delay mechanism for photons, holding them inside the arm long enough for the light to have travelled along the arm hundreds of times, giving more opportunity in a detector for the gravitational wave to interact with the mirrors. This approach was harder to make work practically, which is why Weiss had gone for the simpler setup. Kip Thorne said that Weiss was 'not at all enthusiastic about doing it'. But if the Fabry-Perot cavity *could* be done – and Drever, with his unconstrained enthusiasm, was convinced that it would not be a problem – it would be both more flexible in use and more effective. And this is what was used in the final LIGO design.

Another innovation that Drever brought into the mix was a competitive interaction between Caltech and his home university of Glasgow. Although Drever had a five-year secondment to Caltech, it was on the understanding that his time would be split between the two universities, giving him the chance to spend many hours on plane journeys between the two, thinking up new ideas that he could fire at his teams when he landed. Small prototype interferometers were constructed at both sites to test out the technology, ironing out potential barriers to detection in the way that the mirrors were suspended, in the laser technology and far more.

Drever would fly from one location to the other, spraying out ideas, which the teams would then try to implement. It wasn't conventionally sensible management, but there seems little doubt that the two-site operation did enable a wide range of technical issues to be addressed. In parallel, a team at the Max Planck Institute in Germany was also building a small interferometer, probably the most sophisticated of the early prototypes, and contributing to the collection of information required to take steps towards the incredible sensitivity required by a working gravitational wave detector.

Great though the ideas might have been, the management was not working. After two years of infighting, the National Science Foundation had had enough. The LIGO troika was ordered to replace itself with a single project director who could make decisions effectively. Rochus 'Robbie' Vogt, former chief scientist at the Jet Propulsion Laboratory, was brought on board. It was Vogt who worked endlessly to get the latest estimate of funding approved by the US government – not a trivial task, as the bill for the initial version of LIGO, without any bells and whistles, had now risen to around $200 million.

Entering build phase

There's no doubt that $200 million was a difficult ask in 1991. Not only was the House of Representatives Committee on Science, Space and Technology reluctant to spend such a large sum of money, some of the members were less than enthusiastic on pure science spending in any form. It didn't help that a respected astronomer, Tony Tyson, brought in to testify to the committee against his will, had to admit that all this money could be spent without there being any findings at all. After all, as Tyson pointed out: 'We have perhaps less than a few tenths of a second to perform this measurement. And we don't know if this infinitesimal event will come next month, next year, or perhaps in 30 years.'

Although Tyson is unlikely to have wanted to sabotage LIGO, there was significant opposition from some US astronomers, who weren't used to the kind of big budget projects familiar to particle physicists, and who were suspicious that if LIGO were funded, their own far smaller projects would be at risk (the whole of the annual US astronomy budget at the time was around half that suggested for LIGO). It has been speculated that part of the astronomers' problem was the use of the word 'observatory' in the LIGO name, giving the impression that the physicists were muscling in on the astronomers' territory.

Despite these difficulties – including the usual political wrangling that goes alongside the location of major scientific sites in the US, which in this case ended up with the perhaps not obvious choices of Hanford and Livingston – the funding was eventually passed. Vogt had made it happen, where it is highly unlikely that the troika ever would have succeeded. And yet soon after Vogt was to be fired.

It didn't help that Robbie Vogt was what could be described as a reluctant administrator. He was a scientist through and through, who only tended to take on administrative roles where he felt that the alternative would be to put the position in the hands of someone he felt would be too bureaucratic. Vogt's anti-authoritarian approach was probably far better suited to getting the project the green light than it was to managing the subsequent massive, complex construction project day to day.

It's perhaps ironic that one of the biggest challenges Vogt faced was in dealing with someone even less enthusiastic about outside authority than he was, the Scottish member of the troika, Ron Drever. Vogt appears to have found Drever's highly intuitive and unconventional approach – which as we have seen, undoubtedly delivered some of the key technology for LIGO – impossible to manage. He was not alone in this. Drever resisted attempts to be more systematic and proved difficult for many to work with over his time on the project. After months of rising tension between the two, Drever was removed from his involvement.

Significant though Vogt's contribution was, it was not sufficient to get the LIGO project fully off the ground and operational. While he had been good at getting initial funding and organising the scientific aspects of the project, frustrations with the necessary administrative processes began to show through as discussions began with steel manufacturers and construction companies. Vogt was simply not suited to the bureaucracy required for high-budget business negotiations. He was replaced by Barry Barish, a physicist whose work on building collider technology – the biggest game in town as far as physics spending is concerned – had given him a lot of experience of the kind of large-scale construction that would be necessary for LIGO.

According to Thorne, LIGO got lucky as a result of another major physics project's misfortune. When LIGO ended up looking for the new director, the US had recently begun the construction of the Superconducting Super Collider (SSC). Had this massive project been finished, it would have been well ahead of its European rival, the Large Hadron Collider at CERN, and would, in all probability, have received the laurels for discovering the Higgs boson. But the SSC had just had its funding pulled, which Thorne says left a key player free with perfect timing, and an excellent match for the requirement. For Thorne and the rest of the LIGO team, the news that Barish was suddenly available was anything but bad.

The first job of the new director was to recover the project's position with the National Science Foundation. In recent months, there had been threats to pull the budget – instead, Barish felt that Vogt's estimates had been too optimistic and got the starting budget pushed up to over $300 million. It seemed honesty paid, and he was given the go-ahead. Towards the end of 1994, work started on construction at Hanford, followed by Livingston in 1995.

These sites should have been carefully selected to be away from as many potential sources of vibration as possible, with a flat and open landscape for the 4-kilometre arms of the interferometer. It's common practice these days for traditional observatories to be located far away from sources of interference, such as the Atacama desert in South America. Certainly, each of the chosen LIGO locations was out of the way, but in practice the Livingston site suffered from vibrations from nearby forestry operations, while Hanford had its own distinct problems. The site was able to occupy part of a large area of land that was kept isolated because of a nearby plutonium uranium extraction plant, used to remove plutonium from old fuel rods. This meant that there was a large area kept clear of public activity, though it might seem that the plant itself would cause vibrations.

In 2017, the Hanford LIGO site had to be briefly evacuated when a section of tunnel at the extraction plant, used to house contaminated materials, caved in. Hanford's lead scientist, Mike Landry commented: 'The LIGO sites are indeed selected for well-understood (and generally low) seismic noise, among many other considerations. While no site is perfect, the LIGO Hanford Observatory location is relatively seismically quiet. For example, the PUREX [plutonium uranium extraction] facility is not in operation, is twelve miles distant from the corner station, and provides negligible impact on performance. There are other active facilities on the DOE [Department of Energy] site, like the Vitrification plant, currently under construction. Most of these sites are sufficiently distant to be not particularly impactful.'

Once construction began, one of Barish's key acts was to expand what had been a two-way partnership between Caltech and MIT to take in far more groups around the world. By the time of the 2015 event, the LIGO collaboration involved around 1,200 people working in 80 institutions in sixteen different countries.

The initial version of the two-station observatory went live in 2002. Although there were always hopes that something would be detected during the initial phase, this version of LIGO was expected to fail as a detector – and it did. The whole reason for building the first generation of the devices was to get experience with the technology and at dealing with all the problems that would get in the way of detecting those exquisitely fine vibrations. And in its original configuration, LIGO did just that.

Simulating the universe

A lot of the experimentalists' time was spent during the thirteen years between 2002 and 2015 on ensuring that the equipment was working properly, as it had been gradually upgraded to provide greater and greater sensitivity. But the theoreticians were not left twiddling their thumbs. One problem that gravitational wave astronomers face is that they can't necessarily see using conventional astronomy the sources they observe through gravitational waves. (In fact, none of the successful detections so far has had any light-based evidence to back it up.) So, given a wave trace detected by both observatories, the scientists have to work out exactly what it is that they are seeing.

To an extent this has always been a problem with, for example, radio astronomy, which also involves detecting a wave-like signal in a data stream. But it is much easier to be certain of the direction a radio signal comes from, as most radio telescopes are highly directional, and radio sources are often also visible as light sources in different bands. By combining different frequencies of light from radio through the visible up to X-rays and gamma rays, it is often possible to get a clear picture of what is being observed. But a gravitational wave event may well consist of just a fraction of a second of a wave being detected and that's all the physicists have to work on. No backup. No obvious mechanism for interpretation.

As far back as the 1960s it was realised that the only realistic way to deal with some gravitational wave observations would be to run computer simulations of a whole range of potential sources to produce standard 'templates' that show the wave form expected from such different kinds of event.

As computer technology has evolved in parallel with the development of the project, more and more effort has been put into building comprehensive libraries of the waveforms expected from gravitational wave sources. In the early days, the work was typically done by a handful of people and, according to Kip Thorne, was not going anywhere near quickly enough with this approach. As the first version of LIGO was due to go live, Thorne and his colleagues built up a group of about 30 theoreticians, working full-time to assemble a useable library of simulations to be matched with the signal when and if a gravitational wave was detected.

So, for example, when a pair of black holes spiral into each other and merge, theory predicted that there would be a distinctive pattern where a gradually increasing frequency of wave peaked strongly, to be followed by a series of much smaller, quicker oscillations fading away during the so-called ringdown period, when the merged black holes would be vibrating back and forth as they settled to a single mass. This was the distinctive pattern of the waveform that was detected in September 2015.

Upgrades

Over a decade of operation, and despite any hopes to the contrary, the vast observatories detected absolutely nothing. Kip Thorne suggests that there is a cultural divide between physicists and astronomers which was highlighted by this two-stage process. 'Many eminent astronomers could not imagine that it was a reasonable thing to do to build two generations, where the first generation cost $300 million and the next generation's going to cost you something comparable, to build two generations of instruments before you see anything.'

Others have suggested a slightly different version of history. The physicists may never have truly expected to detect anything with the first generation of the instruments, but this certainly wasn't the impression that they were giving to the funding bodies. According to Harry Collins, a sociologist of science who has been observing gravitational wave experiments ever since Weber's bars, project proposals have always given the impression that the latest generation of the equipment, whatever it is, has the potential to make real detections. In the initial LIGO documents, for example, it was claimed that a range of sources would be within the range of LIGO's first-generation sensitivity, only pointing out, as Collins puts it, 'in the small print' that these hypothetical sources were highly unlikely to exist.

Barry Barish would be succeeded as director by Jay Marx and then David Reitze, who between them would be responsible for getting funding for that second generation of the observatories – the advanced version of LIGO that went operational in the second half of 2015 – and then taking on the extensive rebuild required for advanced LIGO to happen. That would mean a significant extra dip into the money pot. At the time of writing, LIGO has cost around $1.1 billion in total. (For comparison, the construction of the Large Hadron Collider at CERN cost around $4.75 billion, but it also has a much higher running cost than LIGO – so arguably LIGO is still a bargain project in its own way.)

It was not until the pair of sites had received two overhauls that the 2015 event would become possible. In 2009/10 the original configuration was upgraded to enhance LIGO with technology improvements that included a more powerful laser to increase sensitivity. Then, in September 2015, the configuration known as advanced LIGO began to be tested with detectors that were around four times as sensitive as those in the original equipment specification. But even though those working on the earlier configurations of LIGO would never be able to announce that they had found a gravitational wave, there were three occasions when the members of the LIGO collaboration felt a surge of excitement as the earlier incarnation of their instruments appeared to show that the long sought-after signal had been detected.

It's hard to think of another experiment that has generated quite so many dramatic false hopes, especially as some of these disappointments were intentional.

FALSE HOPES 7

It might seem strange that a concept like 'false hopes' could be applied to a scientific discipline. Surely science is all about facts, not something as diffuse as hope? Yet much modern science comes up with probabilistic outcomes rather than definitive answers.

Statistical science

There was a time when science depended on what we could see or touch. This meant that scientists could be factual about *what* they were observing. So, for instance, when Galileo was rolling those balls down carefully constructed wooden slopes to see how they accelerated, there was no question that this is exactly what he was observing. But if he wanted to explain what caused the ball to accelerate, he was limited to theories. So, for instance, when he observed swinging pendulums and decided that the time the pendulum took to make its swing did not depend on how far the bob at the end of the string travelled, it was a theory.

A theory like this could never be a fact, as it would only take one counter-example to negate it. But the more experiments and observations were made, the better the theory could be matched to what actually happened and the stronger that theory became. In the case of Galileo and the pendulum, later researchers discovered that this idea only applied for relatively small swings – it's not true if you make the pendulum travel a long distance – but it's a useful approximation.

As it happens, Galileo was also one of the leaders in a scientific revolution of using instruments that enhanced the senses – specifically, in his case, by making use of the telescope. Where the naked eye saw Jupiter hanging isolated in the sky, Galileo with his telescope was able to assign it a collection of four orbiting moons. The use of such instruments was a huge advance for science, but it also introduced more potential for uncertainty into the observation. When I see something at a distance on the surface of the Earth I might be mistaken about what it is – but I am usually able to go closer and establish what I am actually seeing. I might, for example, see what appears to be a pool of water in the desert, but when I approach it, the water will disappear, as it was a mirage.

However, when I see something at a distance in the night sky, I can't go closer and check what I thought I saw – and the telescope gives more opportunities to see a faint image of something distant and misinterpret it. Around the end of the nineteenth century, for example, a number of astronomers described in detail huge, canal-like structures and greenery on the surface of Mars. These provided ample inspiration for science fiction writers from H.G. Wells to Edgar Rice Burroughs – but they did not exist. It's likely that, peering at something on the verge of visibility through their instruments, the astronomers saw what they *wanted* to see. As we will discover below in 'Going in blind' (page 99), scientists now try as much as possible to overcome the potential for self-delusion.

Modern physics, though, has taken a major step further away from direct observation. Astronomers who peered through a telescope were at least using a human sense to make their observations. But many modern experiments are entirely indirect. Scientists don't observe a phenomenon at all: they see a string of numbers on a computer – or a graphical representation of those numbers – and have to interpret that data, developing ideas on what has caused it. Inevitably there is a degree of uncertainty involved in this process, and science has developed mechanisms to quantify that uncertainty, represented as 'confidence intervals'.

These confidence intervals are often presented mathematically using 'sigmas'. This term refers to the statistical measure of a standard deviation, represented by the lower-case Greek letter sigma (σ). Many natural occurrences, when plotted out on a graph, fall on a normal distribution. This means that the plot fits a bell-shaped curve – with such a distribution of events, around 68 per cent of occurrences will fall within one standard deviation of the mean (average) value, represented by 1 sigma.

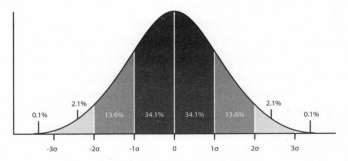

Bell curve showing the percentages of occurrences for different sigma values.

By the time we get to the two standard deviations or 2 sigma level, we have taken in 95 per cent of events. As many as 99.7 per cent fall within three standard deviations or 3 sigma and so on. The higher the confidence interval, the more convincing a result is, because it's very unlikely that an occurrence was purely random.

We ought to be careful here, as even many scientists often get confused over what a confidence interval tells us. Let's say we are testing a hypothesis. We'll take an appropriate one – that what we have observed with our gravitational wave observatory is a gravitational wave and not noise. We do the calculation, and our result comes back with a 95 per cent confidence interval. This doesn't mean that there's a 95 per cent chance that the hypothesis is correct. It means that there's a 95 per cent chance that what we observed would not have happened purely by chance. Or to put it another way, if you did the experiment 100 times, you would expect this result five times with no particular reason for it. This isn't the same thing as a 95 per cent chance of having detected gravitational waves. That would be the probability of the hypothesis being true given the observation, but what we've actually got is the probability of the observation occurring given the hypothesis. These can be widely different.

In the social sciences and psychology, a 95 per cent or 2 sigma confidence interval is often considered sufficient to accept a hypothesis, which causes physicists to more than raise an eyebrow. The problem with accepting such a relatively low probability – we're all familiar with 1 in 20 chances happening – is that it's easy to report inaccurate results. Although physics has its own problems, as demonstrated by the Weber claims, the preference here is for 5 sigma, which instead of a 1 in 20 chance of the outcome happening randomly makes the chance 1 in 3.5 million. Note that for very frequently occurring events this doesn't totally rule out the occurrence. Lotteries are won every week with lower chances of getting a winner. But it does ensure that it is very likely there was a reason for the outcome, and though that reason may not be our hypothesis, it is consistent with it.

This is the kind of measure that had to be applied, for example, when attempting to detect the Higgs boson. No one saw a Higgs boson. They saw a set of numbers in a computer that were consistent with traces caused by other particles that might have decayed from a Higgs – and the chance that this set of numbers would have been produced with no cause, just by chance, was lower than 1 in 3.5 million. The same indirect statistical justification applies to the search for gravitational waves. We know that there were other potential causes of the movement of the mirrors in the LIGO experiment, but should the experimenters get a 5 sigma result, they are saying that the chances are less than 1 in 3.5 million that they would get this data if there wasn't a cause – though their interpretation of that signal is still open to question.

This assessment of probability was achieved, as already mentioned, using the process of 'time slides'. The readings from one of the detectors were slid along the time axis until they coincided with a signal from the other detector. This process was repeated over and over again to obtain a probability of two similar signals occurring at the same time purely randomly. Though in a sense the outcome is faked up, this gives a picture of the background level of coincidences to be able to set that 5 sigma level. For example, the outputs of the two detectors could be moved against each other by a slide of, say, two seconds and the signals compared of the whole run of, say, a week's observing. Then the two-second slide is done again, and the traces are compared once more. Eventually, a picture is built up of the chances of a coincidence occurring over a long period, even though the actual observing period was only a week.

In practice, if a target signal is to be considered a serious contender, it should only happen by chance less frequently than once every 10,000 years – so a large number of these time slides have to be undertaken. For many scientists, this kind of massaging and testing of data is an everyday occurrence, but others are concerned that there may be darker motives behind the manipulation.

Are they real?

There are always conspiracy theorists on the fringe of a discovery who like to pick at big breakthroughs and attempt to show that they are fraudulent. Infamously, there are still people who believe that the Apollo missions never made it to the Moon and that what we saw on our TV screens was mocked up in a studio. None of their assertions hold up to proper scientific scrutiny. The 'evidence' provided by the conspiracy theorists (for example, that the US flag was flying, despite there being no wind to blow it) falls down when the facts are made clear (in this instance, aware that this would happen, the flag's designers provided it with a spring along the top). Perhaps the best counter-argument of all is that a conspiracy like this would simply be too difficult to keep under wraps.

After all, thousands of people would have to be in on a secret like a fake Moon landing. It's inevitable that good evidence of the faking would leak. And the benefits of putting so much effort into fooling the public in these circumstances – especially as there was no attempt to cover up disasters such as Apollo 1 and Apollo 13 – are difficult to comprehend. However, some legitimate scientists have raised questions about whether the results from LIGO are at best an error of judgement and at worst an attempt to mislead.

It is certainly possible to imagine a far more realistic reason for a conspiracy in gravitational wave detection than for the Apollo landings. The conspiracy version would go something like this. A huge amount of money and effort has been invested in LIGO. As is frequently pointed out, over 1,000 people's jobs depend on it and over $1 billion has been spent. Yet until 2015, fifty years of gravitational wave research had produced no results. Not a single thing. How long would the authorities keep pumping good money after bad? Surely, if the advanced version of LIGO had failed, it would be the end of the dream? So, with the best intentions of keeping the science going, the team could have generated a couple of fake detections, just as an insurance policy. That way, the project would continue and would quite possibly get some real detections in the future as the sensitivity of the instruments continued to be increased. Oh, and did someone mention the Nobel Prize?

The reason a conspiracy seems almost plausible is that by comparison with an Apollo conspiracy, it would be so easy to do and to keep covered up. We're not talking about faking up a detailed video of Moon landings – all that would be required is a small change in a waveform over a fraction of a second. Tweaking a relatively small amount of data. What's more, the mechanism existed to do exactly this – it was necessary for the injections that were used to test the system. And it was quite possible for a fake signal to be introduced with only a handful of people knowing about it. Again, the blind injection process used to deliberately produce faked readings to test that procedures were working correctly, could be invoked with only three or four people being aware it had been used.

While there is no suggestion whatsoever that a fake injection did happen, it's not entirely surprising that under the circumstances, some scientists have come out and suggested that the data is being more positively interpreted than it objectively deserves to be. Physicists at the Niels Bohr Institute in Copenhagen and the Institute of High Energy Physics in Beijing published a paper in 2017 suggesting that the 2015 event could have been nothing more than background noise that occurred coincidentally at the two sites.

To get to the impressive-looking waveforms shown as the detection event, the LIGO scientists processed the data to remove large amounts of background noise. What the 2017 paper suggests is that, when that noise itself is examined near to the event, there are correlations between the two detectors – the background noise itself rises and falls in parallel at the two sites – and, crucially, the lag between those rises and falls is the same as the lag detected between the signals identified as gravitational waves. The paper suggests this could be due to signals in nearby calibration lines, which could produce interference that is in step, but with a slight time variation, between the two observatories.

This is a minority opinion. When dealing with this kind of science, dependent on heavy-duty statistical analysis and pre-processing of the data, it's not uncommon that there will be some conflicting opinions. In a response, a LIGO scientist suggested that the correlation in the Danish paper is produced by the mechanism used to analyse the data, involving a technique known as taking a Fourier transform, which breaks down a complex signal into a set of simple waveforms. The LIGO assertion is that the analysis mechanism of the critics was sufficient in itself to produce the apparent correlation without there being any oddities in the data. The opposing scientists disagree and have already cast doubt on the LIGO argument.

At the time of writing, the best-supported theory by far is that the LIGO results are genuine gravitational wave detections, but it is good that the scientific community has not been prepared to leave what could be seen by conspiracy theorists as a suspiciously convenient finding (especially arriving as it did on the 100th anniversary of general relativity) to stand without careful analysis and assessment. However, it is important to be aware that some doubts have been expressed. Gravitational wave events are not clear and simple factual observations, but complex analysis of highly manipulated data. And this also applies to the way the time slides are handled. Even this is not entirely straightforward, thanks to the intervention of little dogs.

Dealing with little dogs

In using the time sliding process, it's possible to get into something of a mess with an occurrence known in the trade as a 'little dog'. (The obscure name is derived from the Big Dog event – see page 105 – which was when this dilemma first came to the attention of the LIGO team.) As we have seen, in the time slide process, the outputs of the two detectors are effectively placed alongside each other, and after each slide, a comparison is made to see if there is a coincidence of waveforms that could have been considered a signal had the waveforms arrived at the same time. But what should be done if one of the two apparent events that line up was the actual event that is being checked using this method? This would mean that the output from one detector is the possibly 'real' event, while the other is definitely background noise.

This combination is known as a little dog, and the confusion arises over what to do with it. If the event that is being tested by this process genuinely *is* real, then it isn't part of the background noise, so shouldn't be included in the time slide process. But if the event isn't real – if it is just the result of a coincidence – then it *is* part of the background and definitely ought to be counted. Unfortunately, the scientists don't know whether or not they consider an event to be real until they have carried out the time slide analysis.

As it happens, the 14 September event was so strong that it was sufficiently different from the background for this not to be an issue, but worrying little dogs would inevitably arise when testing weaker signals. To get around the problem, it was decided that if the event being tested is not so significantly different as the September event was, the background should be calculated including the little dogs, but these should then be removed when looking at any even weaker events in the run, and so on. But this is still only part of the checking: the LIGO experimenters also need to make sure that they are not allowing themselves to consciously or unconsciously influence their data. They need to take an approach known as blinding.

Going in blind

Scientists like to stress how objective they are in their work, but in the end they are human beings with hopes and prejudices, and a good scientific process needs to take this into account. We often now hear in the field of medicine about a drug trial being 'double blind'. For a long time, it has been accepted that we need to keep the people who take the trial drugs in the dark, as they will react differently if they know whether they are getting the real drug or a placebo. But the 'double' part means that the scientists involved in running the trial also do not know what has happened until the results are recorded, as they too are likely to be influenced by expectations. And if they know what is happening they have the potential both of influencing the subject by treating it differently and of being selective with the results.

This double-blind approach of keeping information on what is happening from the experimenter until the data is recorded has gradually spread into other fields such as physics too. It remains far less common than in medicine – apart from anything else, it can be difficult to design an experiment where the physicist is kept in the dark. But without this added protection there is the distinct danger of what is known as experimenter bias.

The classic demonstration of experimenter bias at work dates back to a study at Harvard University in the 1960s. A group of students were asked to perform an experiment on the behaviour of rats, not realising that in reality it was the students who were the true lab rats. The students were given a group of rats and asked to study how well the rats handled a simple T-shaped maze. Some students were given particularly intelligent rats, while others had rats that were considered below average in ability. Each day the students recorded the performance of their rats – and each day, not surprisingly, the intelligent rats far outperformed the dull ones.

As far as those students were concerned, they were doing proper science, objectively recording data that would provide evidence for a hypothesis. In reality, however, they were unconsciously shaping the data to be the way they expected it to be. Because all the rats were identical in ability. There were no especially bright or unusually dim participants.

Over the years, a range of ways has been discovered by which someone can produce this kind of distortion of apparently objective data. The students might sometimes have a borderline case, where a rat sort of solved the maze, but not in the way the experimental design anticipated. For example, a rat might go over a maze wall instead of around it. In making the decision as to whether they should count such an example as a success or not, the students were more generous to the 'intelligent' rats.

Equally they could have resorted to cherry picking. This is a matter of choosing the results that support your hypothesis and discarding the rest. This was not uncommon in the early days of science – Newton, for example, almost certainly did it. Modern scientists are aware that they shouldn't be selective in this way. However, you don't need to be a fraud to cherry pick. Imagine, for example, an 'intelligent' rat failed a test. If there happened to be a loud noise during the experiment, the experimenter might decide not to count that test – to ignore it, because the noise could have distracted the rat, invalidating the result. But had this been a 'dull' rat, the student might well have assumed that the rat would have failed anyway and counted it.

Yet another possibility is that the students unconsciously treated the 'intelligent' rats better than the 'dull' ones, making the 'intelligent' rats more relaxed, less stressed and better able to complete the task. Of course, the LIGO scientists were not going to get a different result by treating a favoured bit of technology better than another. But they did still have the opportunity to cherry pick and make subjective decisions on what to accept and what to reject, which meant that the use of a blinding mechanism seemed essential.

In an experiment like LIGO it's not possible to do the same kind of blinding that is involved in a drugs trial, where the experimenters have no clue about which patients get the drug and which get a placebo. The experimenters have to be able to do some initial analysis on the data to fine-tune the detectors and the software used to process it. As a result, they adopt a practice of splitting the data, having full access to 10 per cent of it for calibration, but keeping the rest hidden until the exact way that it is to be processed is decided upon. It is likely to have been a lack of such blinding that resulted in the unreproducible results from the early days of bar detectors (see page 48).

Even with this blinding in place there were some potential issues. To avoid overwhelming the system with every possible misleading false signal, the LIGO software provides an opportunity to flag up timeframes when the data is likely to be unreliable – for example if a jet flew low over one of the detectors or a major earthquake had produced seismic tremors at both sites. But because the decision whether or not to apply a flag is a subjective one, there is an opportunity for unconscious cherry picking. Similarly, there is human choice involved in the detail of the time slide mechanism.

Time slides provide an objective scientific mechanism. However, there are two aspects of the process that are subjective. Someone has to decide how big the increments are in moving through one log and comparing it to the other. Should the time shift be ten seconds, one second or a tenth of a second, for example? And someone has to decide where to start and stop the comparison process. Bearing in mind random events tend to come in clusters rather than be evenly spread out, it would be entirely possible to select a time period when there are unusually few or unusually many spurious signals.

All of this has had to be taken into account in the many runs of LIGO that have occurred since it first went live in 2002. And nothing demonstrates the risks better than the first serious potential signal detections.

The first alert

Before the 2015 event there were two serious detections, in 2007 and 2010, and one bizarre occurrence in 2004, known as the 'airplane event' that shows an unexpected difficulty arising from the blinding process used. In the set of data used to fix the rules before the blind data was examined, there were no events suggesting that a low-flying plane had caused vibrations. But when the full set was revealed (a process known as opening the box), it seemed very likely that a strong signal was caused by a plane, which was also recorded on one of the audio channels that form part of the background data feed.

Clearly this wasn't a real event (except in the unlikely occurrence that the plane flew over exactly when gravitational waves arrived). But the rules precluded cherry picking after the statistical methods had been fixed. So they either had to use the almost definitely incorrect data in the analysis or to break the rules of fixing the protocol before removing the blinding. As there was no other significant signal present in the whole run, it might seem that it was all wasted effort anyway, but scientists put a lot of significance in data that shows an absence of something. It can help set minimum levels you can expect to observe – providing a kind of useful targeting bracket. But would it make any sense to use data they believed to be nothing to do with gravitational waves?

The group of scientists attempting to decide which way to go with this data could not reach a consensus and ended up taking a vote, which went the way of removing the aircraft data. As sociologist of science Harry Collins, who was present at the vote, observed: 'The procedures of science are meant to be universally compelling to all. [...] It is not that there aren't disputes in science all the time, but putting their resolution to a vote legitimates the idea that they are irresolvable by "scientific" means; neither the force of induction from evidence or deduction from principles can bring everyone to agree.'

The intentional fakes

The aircraft event was an error that arose from the protocol used. However, the LIGO collaboration had always to be aware of the possibility that what appeared to be an event was actually a blind injection. As we saw in Chapter 1, to see how well the systems performed, it was known that at some point during the LIGO observations a fake signal might be introduced to the feed of the two detectors to see if and how it would be spotted. A couple of people are tasked with producing these misleading events – and only they and the project's overall director know that they have happened. For the rest of those involved, the blind injection is a real candidate signal and is treated as such while the teams run through several lengthy steps of the process towards the announcement of a discovery. It is only at the last minute, after months of work, that the truth is revealed.

The blind injection process is important not only to see if what could have been a real signal is picked up by the detection mechanisms, but also to ensure that the time slides, involving moving one set of data against the other in time to create fake coincidences and to give a statistical basis for an occurrence (see page 47), are working properly. It's also probably fair to say that the blind injection process, though accepted as necessary, was deeply disliked by those who were subject to it. It's almost as if they had to sign up to being pranked as part of their job.

As one scientist commented, the possibility that an event was being faked meant that 'All your enthusiasm gets sucked away […] It's messing with your head […]' The blind injection process inevitably reduced the initial excitement of a detection when it was known it could be an intentional fake. Blind injection might be justified, but it calls for a lot of patience.

Equinox and the Big Dog

In September 2007, that patience was put to the test. A signal was detected that passed all the criteria to start the process towards an announcement. The excitement within the LIGO collaboration was extreme. For eighteen months, teams of scientists worked over the data, produced the necessary reports and gradually ticked the boxes on the step-by-step process of confirmation and analysis. It was only then, when they were practically ready to tell the world, that it was revealed that the event was a blind injection. The focus on the 2007 event was so intense that a second blind injection in this period was missed – a real black mark for LIGO procedures. And at the end of the process it all proved to be a drill.

Without doubt this fake detection, known as the Equinox event as it occurred at the autumnal equinox, was frustrating for all involved – but scientists working on this kind of project have to be in it for the long haul. Lessons were learned, both from the missed blind injection and in the complex process of analysis and preparation. It wasn't until the full procedure had been carried out that the teams were able to refine and speed up the way they went about the task.

When the next significant signal was detected in September 2010, that eighteen-month process had been whittled down to just six months. This event appeared to be from the direction of the Canis Major (Big Dog) constellation, hence its nickname. Once more the whole detection procedure was carried out. LIGO sources would later report that the event was 'beautifully consistent with the expected signal' from the merger of a pair of black holes. Alerts were sent out to optical observatories in the hope of confirmation of the event. There was a confirmation from the VIRGO detector, though it was significantly weaker than the LIGO signals. Then the truth was revealed.

It was thought initially that the Big Dog event was a double disaster. After recovering from the disappointment of it not being a gravitational wave, the analysis of the event made while it was assumed to be real was compared with the settings used by the fakers. This should have been a chance to show that the detection process was accurate. But the blind injection should not have looked like a pair of inspiral black holes – nor was it intended to have come from the direction of Canis Major. Luckily, though, it turned out that there was a bug in the program producing the blind injection, accidentally altering the data to produce the results that the analysis had provided.

The blind injections were not just to ensure that the teams spotted the events and analysed the data correctly. They also provided realistic drills in working through the lengthy process from observation to announcement. The hope was that work on a future event could be trimmed down to half the six months it took to process Big Dog. In practice, it would be five months after the September 2015 event that the announcement was made.

Flexing the BICEPs

Although gravitational wave astronomy is largely dependent on interferometers like LIGO, it did seem in 2014 that the technology would be pipped to the post by a totally different piece of kit, known as BICEP2. This was the second in the series of Background Imaging of Cosmic Extragalactic Polarization experiments. Located at the remote Amundsen-Scott South Pole Station, BICEP2 consists of an array of extremely sensitive, supercooled detectors – effectively a specialist collection of small radio telescopes, working in the microwave region, looking for a particular aspect of the cosmic microwave background.

This background radiation permeating the universe and arriving at Earth from all directions has been described as the 'echo of the Big Bang'. If so, it's a very late echo. In the early stages of its development, the universe is thought to have been opaque, as the seething mass of charged particles and energy that made it up would absorb any light generated, rather than letting it through. But after about a third of a billion years, enough of this plasma had converted into normal atoms, and high-energy light was able to blast through the universe. That same light has been going ever since. As the universe has expanded hugely in the intervening period, what was once high-energy gamma rays has (thankfully) been red-shifted, stretching its waves with the expansion all the way down to the microwave region – electromagnetic radiation with wavelengths of the order of centimetres.

This cosmic microwave background radiation has been observed over a number of years, originally picked up by ground-based radio telescopes (it even formed part of the snow that used to appear on analogue TV screens when tuned between stations) and more recently using satellites. However, BICEP2 was searching for a particular form in the background radiation that would add weight to a theory accompanying the main Big Bang theory which has been increasingly challenged of late. Specifically, the experiment observes the polarisation of that background radiation.

Universal stretch marks

The original Big Bang theory developed in the 1930s made a lot of sense. It was derived by imagining running an expanding universe back in time to the point where the area we can potentially see, the 'observable universe' – which now stretches to a volume of space around 90 billion light years across – was smaller and smaller, bringing it back to a point of origin around 13.8 billion years in the past. But this apparently straightforward theory had a few problems.

First, the whole universe seemed to be too uniform. This seems unlikely to us when we look out into space and see a mix of near-vacuum and incredibly dense stars. The universe is anything but uniform on the scale of solar systems or galaxies. But taken as a whole, much of the universe averages out to be of a very similar constitution. To make this possible, it seemed likely that all the points of the visible universe were once in close contact – yet the universe seemed to be too big for there to have been a straightforward simple expansion producing this outcome. Secondly, the structure of galaxies had to have come from somewhere. But where?

In the 1980s, astrophysicist Alan Guth was one of the first to propose the existence of cosmic inflation to explain these problems. The idea was that shortly after the Big Bang, the universe had undergone a sudden and dramatic expansion – far faster than its normal rate. So fast, in fact, that it would have far exceeded the speed of light. (According to relativity, nothing can move within space faster than light, but this doesn't stop space itself from expanding faster.) This inflationary theory suggested that, in a tiny fraction of a second, the universe became thousands of trillions of trillions times bigger. Then the inflation process stopped.

Such a patch to the Big Bang theory does fill in those gaps in its match to reality – both spreading out an early uniformity and transforming tiny quantum variations into the large-scale structures that would act as seeds for galactic clusters. However, there has been little effective evidence for this process happening, and by the time BICEP was constructed there was increasing suspicion in some corners of cosmology and astrophysics that inflation was a failed theory.

The job of the BICEP detectors was to pick up universal 'stretch marks' from the sudden and dramatic impact of inflation. If inflation had indeed happened, it was to be expected that this ripping expansion of space would set off gravitational waves, and these waves should have had an impact on the polarisation of the cosmic background radiation.

The right flavour of light

Polarisation is a property of light that is easiest to imagine if we think of light as a wave. To be precise, as a double wave – a wave of magnetism which creates a wave of electricity, which creates a wave of magnetism, hauling itself along by its own bootstraps, as we saw in Chapter 2. It was the discovery that such a wave could exist only at one speed – the speed of light – that gave the first hints of what light really is. As we have seen, these waves do their wiggling sideways compared to the direction that light travels in, with the electric wave at right angles to the magnetic wave.

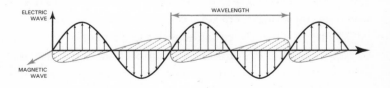

Orientation of waves in vertically polarised light.

Often light consists of a whole mix of polarisations, where the polarisation is defined as the direction of the electric part of the wave. If we think instead of light as a stream of photons, polarisation is a property of the photon that has a specific direction, and each separate photon making up the beam of light can be differently polarised. However, it's equally possible for a light source to emit photons all polarised in the same direction, in which case the light beam as a whole is described as being polarised – for example, a laser's light is polarised – while some interactions between ordinary light and matter cause a higher than usual quantity of the photons to be similarly polarised. This is the reason that, for example, reflected light is partially polarised, which is why Polaroid sunglasses, which only let through light polarised in one direction, are able to filter out a considerable amount of reflections.

Gravitational waves caused by the inflation of the early universe were expected to produce a pattern in the polarisation of the first light to cross the universe – and it was this polarisation that the BICEP2 detector (and its predecessor, BICEP1) were set up to study. Not only would an appropriate detection give support to the inflationary theory, but depending on the strength of the gravitational waves and their impact it should have been possible to distinguish between a number of variants in inflationary theory. And, of course, this would be evidence for the existence of gravitational waves, which were yet to be directly discovered.

There was, then, considerable excitement when on 17 March 2014 it was announced that BICEP2 had detected exactly the type of polarisation in the cosmic background radiation expected from these 'primordial' gravitational waves.

Smoke gets in your eyes

For a brief period of time, BICEP2 was considered to be a triumph. But over the following months, as more detail emerged, it was suggested that there was a problem with the findings, and by September 2014, data from the Planck satellite had made it clear that the results were nothing to do with polarisation in the cosmic background radiation but were instead caused by interstellar dust.

It didn't help that the BICEP2 team decided to announce their results at a press conference at Harvard University before the kind of scrutiny of a scientific paper that the peer review process allows. Within hours of the initial announcement there was a groundswell of doubt from areas of the scientific community. The Harvard team behind BICEP2 had been aware of the potential effects of dust, but they had miscalculated the impact that the dust would have, significantly underestimating its contribution to the signal. Within weeks of the press conference, Paul Steinhardt from the Center for Theoretical Physics at Princeton University was writing in *Nature* that 'Serious flaws in the analysis have been revealed that transform the sure detection into no detection.'

This problem reflects the difficulty inherent in this kind of project. Very small variations in detections of this microwave radiation are being ascribed to particular causes – but it's all too easy for something as simple as varying amounts of dust getting in the way of the signal and scattering the radiation to totally confuse the observation.

The BICEP team have not given up the search – a more sensitive BICEP3 is in operation at the time of writing, but as yet we still have no evidence of these primordial gravitational waves.

Not the real thing

Even if the BICEP2 detection had been genuine, it would have been far less significant than a successful LIGO observation. We're back here to a form of indirect observation of the impact of gravitational waves. BICEP2 was looking for the effect the waves would theoretically have had on the cosmic microwave background, rather than making a direct observation of gravitational waves that actually still exist, and providing a whole new vehicle for astronomy. The observation of the hoped-for variation in the polarisation of the cosmic microwave background might have given useful clues to some of the early development of the universe – notably to add or subtract weight from the inflation theory – but by comparison with LIGO it would have been a minor advance.

Unfortunately, though, all the evidence was that LIGO too would be a failure, with the only detections proving to be either errors or blind injections. Until 14 September 2015.

THE GREAT WAVE

8

Realisation revisited

For the first two days after 14 September 2015, emails flying around the LIGO collaboration were tentative as more information gradually became available. This was the third time that a signal had been detected in the engineering runs of advanced LIGO, but this was only a test phase, and glitches were to be expected.

The earlier engineering run 'signals' had been easily dismissed as noise. But by lunchtime on 16 September, the Burst group asked the consortium to initiate 'Step 1 in the process of claiming a first detection'. The Burst group was the team responsible for the detection system that flagged up the signal initially, looking not for matches to the many models of cosmic activity but simply a strong, unexpected waveform. At the time, the 'false alarm rate' (see page 119 for more on this) was given as better than once in 200 years, an alternative way to the more common 'sigma' approach of describing the significance of data.

No one had expected to see a promising candidate signal so soon. This was the advanced version of LIGO running properly (though not at full detection capability) for the first time. But the LIGO observatories were temperamental instruments at the best of times, spending around half their potential working time out of action, either because there was a fault in one of the systems or due to environmental impact from external vibration. It seemed ridiculously lucky to find a signal that could be a gravitational wave so soon – and a strong signal at that. Kip Thorne has said that they never expected the first detected signal to be obvious, but thought they would have to painstakingly extract it from background noise before it became noticeable. This event was a clear wave pattern, though, visible with only the minimum of pre-processing.

That's not to say that the detectors were silent until the gravitational wave arrived, then suddenly burst into activity. With the immense sensitivity of the interferometers, there is always something being recorded, a wavering trickle of background noise. But what is striking is that for the short period of the signal, lasting only around 0.07 seconds, there was both an initial increase in intensity and a sudden coming together of the patterns from Hanford and Livingston. Where usually the signals received at the two detectors move randomly in an unconnected manner, for this brief period, the two were in synch like a pair of dancers who had finally mastered their art.

It ought to be stressed that even here there *was* some manipulation of the signals before they could be compared. This is a standard set of manipulations, primarily to cope with the difference in orientation and construction of the two observatories. The two arms at each observatory are oriented so when the 'left' arm is being squashed at Hanford it's the right arm that is squashed at Livingston, and vice versa, so one of the waveforms has to be turned upside down. There are also more subtle differences to allow for the curvature of the Earth and for the subtle variations in behaviour between the two observatories – as Thorne has put it, the teams had to learn the 'personalities' of the different detectors. These are not perfectly reproduced, mass-produced devices; each is a one-off construction.

The 'Step 1' request was the start of a potential four-stage process of careful checking and preparation for announcing the first gravitational wave detection to the world. One advantage of having a project that takes decades to complete is that those involved have time to be highly prepared for what they will do if it looks likely that they have a breakthrough. The process had only been started fully twice before, for the Equinox and Big Dog events described in the previous chapter.

Baby steps

Once the procedure was started, there was more than a little concern among the teams that once more they could be facing a blind injection. Although it had rapidly been established that this event wasn't an engineering injection – a simple, publicised test of the systems – it had been agreed to continue the blind injection process that had produced the Equinox and Big Dog events. This was, to say the least, not a universally popular decision. Some felt that the lessons had already been learned and that blind injection was a matter for the testing phase of the project, out of place during real observations. It was even said that on past events, some scientists had cheated and sneaked a peek at the data that would confirm the use of a blind injection to avoid wasting their time on unnecessary analysis.

Conventional astronomers didn't have to put up with this kind of interference in their work. Blind injections had been important for the development of the systems, but they didn't get you a write-up in *Nature*. Yet they had been kept in the programme for LIGO, arguably because they allowed for plausible deniability when it came to information leaks. One problem with a detection is that it requires months of work to come to a decision. This means that it can be very difficult to keep the discovery quiet, especially in a connected, social-media-savvy world. The LIGO hierarchy would have been aware of the too-early announcement of the BICEP2 results in 2014 and would not have wanted to repeat that embarrassment. The possibility of a blind injection meant that, should it get out that the collaboration was in serious analysis mode, the PR message could be 'Nothing to see here – probably a blind injection.'

The one consolation for the anti-blind injection camp was that this was an engineering run, and so there was no apparent justification for having a blind injection during observations that were about testing the functioning of the machinery and making last-minute tweaks to the setup, rather than testing the analysis process. But this logic wasn't enough to convince everyone. Even though at least one member of the senior management announced to the collaboration that there definitely were no blind injections involved, the impact of the previous faked events was psychologically significant: a good number of the scientists remained suspicious. They even considered the possibility that hackers had maliciously injected data into their systems – though it seems highly unlikely that outsiders would have had sufficient knowledge to make a suitably deceptive signal.

This wasn't the only concern facing the teams who were analysing the data. Because the signal had occurred during an engineering run, there was only a limited amount of data available between changes to the system which could be used to do the statistical time slides to look for chance occurrences. Not enough, in fact, to verify the data to the required level. The question arose as to whether data from the first proper observation run, started four days after the detection, could also be used as background material for checking. Whether or not this was possible depended on how similar the setup of the equipment was for the observation run. If it had changed significantly from the engineering run – which would normally be the case – it would invalidate the results.

Making it possible to carry on the time slides into the observation run caused significant problems for the operational teams. The whole point of the engineering runs was to be able to tweak the system and to fix any minor technical problems. But if any changes had been made to the experimental setup, it would not be possible to make comparisons between data from the engineering run and the observing run. This meant as much as possible 'freezing' the detectors as they were until sufficient data had been gathered. This freeze would last from the start of the first observing run on 18 September all the way through to 20 October. There was grumbling from some of the scientists and engineers, as known faults had to be tolerated.

To add to these concerns, this could not be considered a fully blind search in the usual sense, as the signal had been discovered in the open data. However, a large proportion of the data could still be kept blind, meaning that the whole process wasn't being fine-tuned to the data, and with such a clear signal it's arguable that the blinding process was less necessary. As long as, for example, the protocols for the time slides had been agreed before any analysis had been made, there seemed to be limited opportunity for things to go wrong.

Meanwhile, there was considerable discussion over just how long these time slides should be. Make them a lot shorter and you get many more comparisons from the same amount of data – but at the same time, make them too short and it's possible that the same burst of vibration could be compared twice, because the apparent signal lasted longer than the time slide.

Opening the box

When around a week of the first observation run data had been collected, it was suggested that it was time to 'open the box' of the blind data and commit to time slides. At this stage, there wasn't enough data to get to the required significance levels for publication, but there was enough for the collaboration to overcome any internal doubts. The moment came on 5 October. Despite concerns about potential double counting, it had been decided to use time slides of 0.2 seconds, providing ten or fifteen times as many comparisons as would have been the case with a two- or three-second slide respectively. The process had already been done before the 'box' was opened (but the outcome had not been examined), meaning that the result would be known almost immediately.

If there had been any matches in the time slides that also fitted one of the simulated waveforms, then the discovery could not be given a high enough significance of not being a random occurrence. But the results showed that there was no such match. The only match was the actual 15 September event, which was given an expected false alarm rate of about 1 in 30,000 years. It might seem impossible to deduce what would happen over 30,000 years from a few days of data, repeatedly compared – what the false alarm rate tells us is that *if things had continued as they were during the sixteen days covered by the time slides with no changes*, we would expect around 30,000 years to elapse between false alarms.

Strictly speaking, at this stage the event could still have been a blind injection. The envelope revealing whether this is the case is not opened until step four of the process, just before going public. But the feeling in the collaboration was increasingly that this was the real thing. After the disappointments of Equinox and Big Dog, excitement was growing.

Preparing for the world

Perhaps surprisingly, as the teams prepared their first paper and announcement there was heavy debate on what it was that they would be announcing. Two key claims, and how to word them, would cause contention. These were that this was the first direct detection of gravitational waves and that it was the first direct observation of black holes.

While there is no doubt that the scientists involved wanted to be the first with a big breakthrough, they were aware that gravitational waves had already been indirectly detected and some were wary of claiming too much for

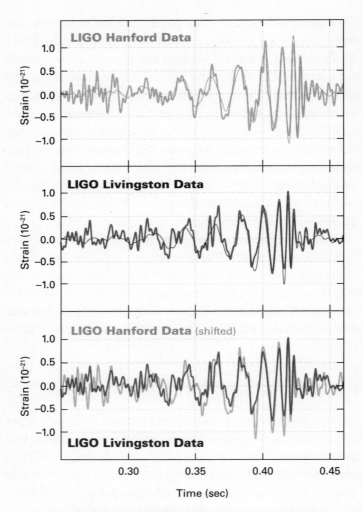

**September 2015 event data from Hanford (top),
Livingston (middle) and combined (bottom). Faint
line on top two shows the pattern predicted
by the model of inspiral black holes.**

LIGO

themselves. Self-aggrandisement is rarely considered a good thing in science. If the source of the waves definitely was a pair of inspiral black holes, the claim to have made the first direct observation of these cosmic phenomena would definitely be true. All previous material on black holes has either been from theory or from their impact on their surroundings, where a black hole seems the most likely culprit. But some of those in the collaboration pointed out that, in principle, the cause could be some other cosmic phenomenon.

The source was assumed to be black holes because theory predicted that black holes would behave this way and would produce exactly this kind of waveform. But the theory could be incorrect, or there could be another phenomenon that produced the same effect. Think, for example, what you might induce from hearing a large crash outside your house without looking outside. You might reasonably think that the noise was caused by a road accident. However, that same sound *could* have been caused by something else – an engine falling off a plane, for example. If you relied on your theory alone, it would be too much to claim that you had directly detected a road accident.

Such was the level of dispute that the team producing the paper took the unusual step of a poll among the collaboration to decide on a number of elements of the way the discovery should be presented. The results came back deciding the title of the paper should not contain the word 'direct' (or refer to LIGO), though by now, resistance to being definitive about black holes had been overcome and all the options referred to a 'binary black hole merger'. However, a second question on the poll got the answer that it was acceptable to use the word 'direct' in the body of the paper.

In the end, the paper would be entitled 'Observation of Gravitational Waves from a Binary Black Hole Merger', while both the abstract and the conclusion contained the sentence: 'This is the first direct detection of gravitational waves.'

Secrets and lies

Perhaps the most controversial aspect of the LIGO discovery was the way that secrecy was handled. It's very difficult when working with such a large consortium to keep a possible discovery entirely under wraps, but initially the LIGO hierarchy did not want to risk the kind of early announcement disappointment that had happened to BICEP2. Scientists are also always wary that if they share data too early, others might be able to do an analysis of it faster than they can – and win the accolades.

This has certainly been a problem in the past, even with something as abstract as a theory. When Albert Einstein first came up with the basics of the general theory of relativity, the theory that would eventually forecast the existence of gravitational waves, he had shared some of his early work with the leading German mathematician David Hilbert. As a result, Hilbert – without doubt a better mathematician than Einstein – set out to develop his own gravitational field equations, and only failed to beat Einstein to publication because he made a last-minute mistake.

However, despite this urge to keep things under wraps until a clean announcement can be made, we expect honesty from scientists who are funded by public money. There should be a clear line between keeping a discovery secret and lying about it. In practice, though, the attempts to cover up the detection began badly and got worse. Just a week after the discovery, the LIGO media spokesperson issued collaboration members with instructions on how to deal with external queries, including: 'We heard you sent a [gravitational wave] trigger to astronomers already – is that true?' The suggested answer included the downright misleading phrase 'we have been practising communication with astronomers in [the last engineering run]'. What was going on in reality was no drill.

Despite the attempted cover-up there were almost inevitably going to be leaks, given the number of people involved. As early as 25 September, the physicist and science communicator Lawrence Krauss had tweeted about a 'rumor of a gravitational wave detection'. But on the whole, there was a lack of media excitement as the idea that this could just be another blind injection was widely spread by LIGO public relations.

The cover-ups would continue to be compounded, as two other detections occurred during the period of time when the teams were preparing to go public. One event, on 26 December, was a reasonably strong signal, while an earlier one on 12 October, which was too weak to be definitive, was eventually given an 80 per cent chance of being a real detection. Although the teams knew about these events when they announced the 14 September discovery, and the existence of them made it much more likely that the first detection was genuine rather than a random error, the new detections were kept quiet by the collaboration, still intent on secrecy. While there's an element of sense in keeping things quiet for a first discovery before announcement, it seems odd that this was felt necessary.

As the announcement got closer, more rumours were spread on social media, getting picked up in mainstream media. The collaboration worried about what to do about alerting the conventional astronomers to the 26 December event. The usual way to do this was simply to flag up the event in the large shared database of all promising signals. This had been avoided for the September event by using a different mechanism, but it could not be done too often, so the data was toned down – in effect, the information provided to astronomers was faked to avoid it being such an obvious detection.

Even at the announcement of the September 2015 event, the deception continued. By the time of the press conference on 11 February 2016, the members of the LIGO collaboration were well aware that there was one definite and one probable event in the more recent data, but the party line was that they couldn't comment on the observing run, as its data had not yet been analysed. In the end, while it's fair to consider the knowledge of the detection to be the property of the consortium, it is perhaps unfortunate that this deception was taken to such lengths.

Einstein proved right, or wrong?

When the news was finally made public in February 2016 there was a brief but strong spike of worldwide interest, as the discovery was splashed across many front pages. However, the gravitational wave detection did not get the same level of public response as did the Higgs boson discovery at CERN in 2012. It's arguable that this was because the media mostly seemed to misunderstand what the LIGO work was all about, and its significance. The headlines mostly trumpeted that Einstein had been proved right. Yet this was both technically incorrect and a massive understatement.

While it's true that this discovery did support the general theory of relativity, and Einstein's prediction that there should *be* gravitational waves, neither of these was in any real doubt. General relativity had been confirmed many, many times by all kinds of means – the everyday technology of satellite navigation still depends on correcting for it. And Hulse and Taylor's 1970s observations of the neutron star duo had provided evidence for the existence of gravitational waves. If anything, the headlines got it back to front, as Einstein had in reality predicted that gravitational waves would never be detected – so the discovery proved him wrong, rather than right. But what most of the press announcements missed was the far greater significance of the detection: that this was the first major new way of doing astronomy since the invention of the telescope.

It's true that telescopes have come on a long way since the end of the sixteenth century when they seem to have been invented (no one is quite sure when they were first used, or by whom). Their observational spectrum has been vastly expanded with the addition of everything from radio to gamma rays. But all telescopes have depended on light to provide a view of the universe. LIGO, by contrast, was using a whole different type of wave to look out and observe things that had never been observed before.

Take the 15 September event. It had proved a remarkably good match to the model's prediction for the behaviour of two black holes, their orbits decaying so that they spiralled into each other. The two bodies of around 36 and 29 solar masses merged with a vibration that gave off around three solar masses-worth of energy in the form of gravitational waves before a rapid 'ringdown' as they settled into a single, larger black hole. Those waves travelled from their distant galaxy around 1.4 billion light years away at the speed of light, arriving at the Milky Way around 50,000 years ago, before any records of human history exist. On 14 September 2015, they first hit the Earth somewhere around the Antarctic, travelling up through the planet to register on the Hanford and Livingston detectors.

This was far more than 'Einstein proved right' – it was a whole new branch of astronomy being founded. And for the future, it could result in many more findings and the ability to test out theories that to date have proved impossible to confirm or deny.

LOOKING TO THE FUTURE 9

The LIGO events of 2015 began with a fluke, but soon provided our first, transformational view of the universe through the medium of gravitational waves. However, these observations were only the start. LIGO has been operating in conjunction with the GEO600 detector in Germany in its second observational run since 30 November 2016. And the current incarnation of LIGO and its partner sites are just the beginning for gravitational wave astronomy.

Further detections

At the time of writing (October 2017) there have been what are coyly described as 5.8 detections of gravitational waves. There were two solid detections in 2015 – the original September event, still by far the strongest to be found, and the second on 26 December. There was also a third event in between the two on 12 October, which Kip Thorne describes as '0.8 of a detection' as it is less certain than the others but is now indeed considered to have an 80 per cent chance of being the real deal.

On 1 June 2017, a third solid event was announced, dating back to 4 January of that year. Like all the previous observations, this appears to have been a pair of inspiral black holes, recorded as their merger sent shockwaves through spacetime. The new detection neatly fills in the gap between the other two mergers. The first event produced a final black hole with around 62 times the mass of the Sun, while the event on 26 December resulted in a far smaller 21-Sun mass. The new observation sits at about 49 times the mass of the Sun for the final black hole.

As always, for the event announced in June 2017, there was a tiny delay between the signal being detected at the two observatories, arriving around 3 milliseconds earlier at Hanford than it did at Livingston, due to the direction from which the wave originated. Bruce Allen, one of the directors of the Max Planck Institute for Gravitational Physics in Hannover, commented: 'With another event of this kind, we are realizing that heavy binary black holes are more common than we had believed just a little over a year ago. A lot remains to be learned – this is an exciting time for the new era of gravitational-wave astrophysics!'

This merger of black holes with original masses of around 31 and 19 solar masses took place around 3 billion light years from Earth, around twice as far away as the original September 2015 detection.* There was an element of luck in the detection, as the automated system at Hanford,

---

* The distance calculation for the event is based on the expected strength of the wave at the source compared with the strength when received on Earth.

looking for candidate signals, had an incorrect setting, but a post-doc researcher at Hannover, Alexander Nitz, was visually checking candidate signals and picked out this event, initially from the Livingston data, then finding the corresponding event at Hanford.

As of October 2017, there were five candidates that could be significant in the data from the observing run that finished on 25 August 2017. The first of these to be confirmed, a fourth black hole merger, added the VIRGO observatory to make a more accurate three-way observation. In mid-October this was joined by the first detected neutron star collision, producing a distinctive two-minute long signal. Conventional astronomy backed this up with everything from radio to a gamma ray burst.

A third observing run, with even higher sensitivity, is due to be started in late 2018.

A hole in what?

Black holes occupy a strange position in cultural awareness. The term 'black hole' has become a common usage for a kind of mysterious bottomless pit ('there was a black hole in the finances'), and from portrayals in science fiction movies such as *The Black Hole* we have got a strange, distorted view of what is involved. Even so, the term has become sufficiently common that it has been possible to get this far through the book without bothering to say what a black hole is. However, it is important to go into these remarkable bodies in a bit more depth, because until the 2015 direct observation was made, for all we knew they might not even have existed outside of fiction.

It is certainly possible for an obscure theoretical concept from physics that almost certainly doesn't exist to become well-known, mostly due to science fiction, to the extent that many assume it is an accepted part of reality. Take wormholes in space. Like black holes, these are hypothetical constructs based on the general theory of relativity. A wormhole, also known as an Einstein-Rosen bridge, links two points in spacetime that can be widely separated in normal space. But if it were possible to pass through the wormhole, a traveller would be capable of getting from one point to the other almost instantly.

Wormholes crop up time after time in fiction as a way to get around the vast distance involved in interstellar travel without travelling faster than light. And, as an intriguing theoretical model, they have been explored at length by physicists. But it ought to be stressed that no one has ever seen or made a wormhole, nor is there any evidence for the existence of a real wormhole out in space.

While it is theoretically possible to imagine how a wormhole could be constructed, it effectively involves somehow linking together a black hole (of which more in a moment) and a white hole, which is effectively an anti-black hole. Where nothing can get out of a black hole, nothing can get into a white hole. And even though such a construct could exist in theory, we know that if you did have a wormhole and tried to travel through it, it would immediately collapse. It could only be kept open if you had large amounts of another hypothetical construct, negative energy.

For a surprisingly long time, black holes occupied a similarly theoretical position to that of wormholes, though they have a longer history. Black holes were conceptually 'invented' long before the general theory of relativity came along, and the possibility they *could* exist was one of the first implications to be deduced from the general theory, though no one (certainly not Einstein) took this seriously.

The first suggestion of how a black hole (or, as it was then called, a dark star) might exist came not in the early years of the twentieth century, but in the eighteenth. The English astronomer John Michell, born way back in 1724, realised that if a star's escape velocity was high enough, light would never escape. On Earth the escape velocity is about 11.25 kilometres per second. If I throw a ball up slower than this it will fall back to Earth. Throw it up faster than the escape velocity and it will get away from the planet's gravitational field before that field has slowed it to a stop. So, if a star were so massive that the escape velocity was higher than the speed of light, Michell argued, light could not escape and the star would be dark.

This was just wild speculation without any theoretical basis, but in 1916, just a year after Einstein published the equations for the general theory of relativity, a German physicist called Karl Schwarzschild came up with a solution to those equations for the case of a non-rotating spherical body – a simplified model of a star. At the time, Schwarzschild was fighting in the First World War trenches, but somehow was able to abstract himself from the horror sufficiently to work with the complex mathematics. One interesting possibility that emerged from his model of a star was that there was a special limiting size, now called the Schwarzschild radius, which is $2GM/c^2$ where G is Newton's gravitational constant, M the mass of the star and c the speed of light.

If a star turned out to be smaller than the Schwarzschild radius for its particular mass, then that distance away from its centre would form a strange kind of spherical boundary called the event horizon. This was the point of no return – anything within the event horizon, including light, would never be able to escape. At the time, no one envisaged that this would actually happen, because stars are much bigger than the Schwarzschild radius for any particular mass, but it was an interesting theoretical oddity.

It was only as quantum theory developed and physicists got a better understanding of how stars function that it was realised that stars go through an evolutionary lifecycle, changing throughout their lifetime as lighter elements fuse to make heavier ones. In some cases, at the ends of their lives, it seemed possible that stars could either form a neutron star, or, if the gravitational force was strong enough, overcome the force of last resistance, the Pauli exclusion principle. If that happened, there should be no stopping the collapse under gravity, although the resultant 'singularity' would have no dimensions, making it infinitely dense – which suggested that the physics of the time (and of the present, for that matter) could not accurately describe what would actually occur.

Schwarzschild, whose name appropriately meant 'black shield', did not call this hypothetical star a black hole. In fact, we don't know who did. The American physicist John Wheeler, who was the young Kip Thorne's thesis supervisor, certainly popularised the term, first using it in 1967, but it had already been used three years before by an unnamed commentator at an American Association for the Advancement of Science meeting. Whoever named it, though, the black hole was still a controversial, and quite possibly fictional concept in the physics community when gravitational wave research began.

Over time, a huge amount of effort was put into developing mathematical models of how black holes would behave, if they *did* exist. These models resulted in whole careers where theoreticians, most famously Stephen Hawking, developed complex concepts for the way black holes would interact with matter, virtual particles, light and more. There was entertaining speculation about what a human would experience if he or she fell into a black hole, introducing the term 'spaghettification' for the stretching out of a body to form a long, pink strand due to tidal effects. Because of the infinite extent of the singularity, some even suggested that a black hole could be a gateway to another universe, though it was never clear how the traveller was meant to emerge.

All this time, fascinating though the mathematical constructs were to those involved, the black hole could have been pure fantasy. A surprising amount of modern physics is more about creating elegant mathematical models of what might be than about observing reality, and there remained no direct confirmation of a black hole's existence. But gradually, indirect evidence began to emerge. By definition a black hole is not the kind of star you can see in the sky, because no light comes out of it. But if one existed, it would have an effect on anything nearby.

It's worth stressing that this effect is not in any way like the 'super vacuum cleaner' shown in bad science fiction. If you were orbiting a star as it became a black hole, and somehow managed to avoid the debris and radiation that would be hurled around, you would continue onwards in a steady orbit. At any particular distance from the centre of the collapsed star, its gravitational pull would be no more than it was before the black hole formed. However, the big difference would be that you could now get far closer to the star, because the same amount of mass would be compressed into a much, much smaller volume (theoretically an infinitely small volume, but from the observer's viewpoint, the insubstantial 'surface' of the black hole would be its event horizon).

This heavy-duty pull when close to the black hole would mean that any gas and dust that ventured close to the event horizon would be accelerated dramatically towards it. This acceleration would be so strong near to the event horizon that dust and gas would glow incredibly brightly as it rushed inwards, blasting out high-energy light. So, though a black hole itself would be invisible, we would expect to see its impact on the environment around it. And as telescopes got better, they were able to pick out a good number of examples where this kind of activity seemed to be taking place.

In some cases, it was the in-falling surrounding material that could be seen. In others, the suspected black hole appeared to be stripping off outer layers of a close-orbiting binary star. And at the centre of galaxies, including our own Milky Way, there appeared to be vast black holes with masses of millions of times that of the Sun. It has even been suggested that these supermassive black holes are effectively the seeds that enable galaxies to form in the first place.

By the time LIGO became operational, then, a vast amount of effort had been put into predicting the behaviour of hypothetical black holes. And many observations had been made of effects that could have been produced by these remarkable bodies. This is one reason why the LIGO result was so special. This new kind of astronomy meant we could detect waves emanating directly from one of the most mysterious and fascinating entities in the universe. For the first time it was possible to directly detect a black hole.

Black hole bonanza

At the time of the detection, although there were plenty of simulations of black hole binaries (pairs of black holes orbiting each other) spiralling into each other, it wasn't certain whether any such collisions could be observed. This is because orbiting black holes would have a huge amount of momentum to gradually lose before the final moments of the inspiral, and it was only in those final few seconds that gravitational waves strong enough to be detected would be emitted. Some even suggested that the universe was too young for *any* pairs of black holes to have reached this state. Now, though, this possibility can be ruled out – and with at least four already detected, they seem far more common than was first expected.

The biggest problem in understanding black hole binaries is working out how they managed to come into being in the first place. A black hole is the remains of a large star that has collapsed. But to form a binary that could decay so much that a pair of holes merged with each other in the lifetime of the universe would require the black holes to be formed very close together. Thirty million kilometres, around a fifth of the Earth's distance from the Sun, would be the maximum distance they could start off apart. But bearing in mind that to form a black hole, a star usually blows off a large part of its external material, that original star would typically be significantly larger than this maximum orbital distance. It seems as if the stars forming the black holes would have to be touching – which doesn't make sense.

We still don't know for certain how these black hole pairs could form, but the gravitational wave group at Birmingham University has come up with one mechanism that would make it possible. They suggest that a pair of large stars were initially orbiting each other, much further apart than the black holes end up. During the later phases of their lifetime in which they expand in size, material would flow between the stars in a process known as mass transfer. The Birmingham group has suggested that if this material ends up as a large envelope of hydrogen around the two stars it could be blown off, blasting the now significantly smaller stars closer together, starting the process that would end up with the black holes merging billions of years later. For this process to work, the stars would have to be almost entirely made up of hydrogen and helium, making them members of the first generation of stars after the Big Bang.

The initial observations by LIGO have given us far more confidence that black holes exist. This is a remarkable step forward in its own right – but future observations give us the potential of testing out some of the huge amount of theory that has been built up about black holes over the years – some of which has, until now, seemed to be purely hypothetical work that would never be more than an entertaining speculation, as it was assumed that it could never be tested out without visiting a black hole many light years away.

For example, LISA (see below) should be able to use the waves generated by a small black hole orbiting a supermassive black hole to map out the spacetime geometry of the bigger black hole for the first time, to see if the effect of such a dense body on spacetime matches the predictions of theory. Another possibility is to confirm or dismiss a theoretical alternative to a conventional black hole structure called a naked singularity. If one of these were orbited by a star, then the pattern produced would be markedly different from that when a star orbited a conventional black hole.

The naked singularity orbits would be chaotic in the mathematical sense, tracing complex paths – so should naked singularities exist, it would be possible to distinguish them from conventional black holes.

If naked singularities were ever found it would have a major impact on our understanding of general relativity, which would need significant modification. Until recently, their existence has seemed highly unlikely, as no predictions of their existence had been made for universes with fewer than five dimensions. But in 2017, physicists at Cambridge's Department of Applied Mathematics and Theoretical Physics found a theoretical way for a naked singularity to exist in a conventional four-dimensional universe (three of space + one of time) like our own – however, no one is holding their breath, as current theory suggests that such singularities could not exist in a universe in which, as with ours, there are charged particles.

Spanning the world

One significant challenge that has to be faced when moving on to the next version of LIGO from the 2015 configuration is that it will no longer be possible to ignore quantum mechanical issues. When we study the behaviour of very small objects, such as atoms, quantum physics reigns supreme. A fundamental aspect of quantum theory is that most properties of a quantum object are probabilistic, not having a specific value until the object has a direct interaction with something else. So, for instance, until it is measured, a quantum particle doesn't have a precise location – there is a degree of uncertainty attached to it. That uncertainty can be precisely calculated, but we can't say exactly where we're going to find that particle.

The measurements used in the various versions of LIGO up to advanced LIGO were just about large enough not to have to worry too much about quantum behaviour of the particles in a solid, which are more constrained than, say, those in a gas. However, with the next upgrade of the equipment, due around 2020, the physicists will need to take into account this uncertainty in the location of the atoms that absorb and re-emit photons of light in the process of reflection at the mirrors. This means that the distance between the mirrors cannot be specified exactly. This isn't a project-killer – the uncertainty could be accounted for in the calculations – but it is something that the experimenters will have to build into their models. The mechanism to be used is given the impressive-sounding name 'quantum non-demolition' and effectively eliminates the quantum error by measuring two different properties and eliminating the common factor that comes from quantum uncertainty.

As the sensitivity of the system is improved, meaning that LIGO can both see further and detect less energetic gravitational waves, it is expected that the number of events will go up considerably. At the time of writing, LIGO is detecting around one possible black hole merger event every other month. But when the system reaches its full design sensitivity – expected to be in 2019/20 – it should be able to see around three times further than it currently can. This means that it will have access to about a 30 times greater volume than it has at present: the expectation is that a few black hole inspiral events a week will be detected.

With increased sensitivity it is expected that other types of source will come within LIGO's reach: pulsars (see page 57), binaries of a black hole plus a neutron star, and neutron star binaries. It's also possible that the gravitational waves produced by the exploding star that is a supernova will be detected, though these are relatively rare. As with the neutron star collision, there is every likelihood that there will also be light-based detection accompanying the gravitational waves, allowing for far more information to be captured about the event, and for confirmation of the effectiveness of the computer models used to identify events.

To make a practical link-up with more conventional astronomy, gravitational wave astronomers have to have a decent handle on the direction the gravitational waves have come from – and here there is a problem. Although the distance between Hanford and Livingston is enough to give a degree of directional information, a detection that is limited to these two observatories can have a range of sources spanning a wide latitudinal segment of the sky. So, for example, the first signal in 2015 could have originated anywhere between overhead at the South Pole and above South America.

This is where other observatories can add to the mix – a worldwide network of detectors would make it possible to pinpoint the source of the waves far more accurately. Just as a satnav pinpoints its location from the difference in time for signals to reach it from several GPS satellites, so by comparing the time that gravitational waves arrive at a network of observatories, it becomes possible to home in on a specific direction. The first major addition to the network was the advanced version of VIRGO near Pisa in Italy, which went live in 2017. A third LIGO observatory, originally intended for Australia, but now to be located in India after funding issues, is currently boxed up and ready to go, though it will be at least 2020 before this is ready to make a contribution. Finally, a significant new Japanese gravitational wave observatory is being constructed in the Kamiokande mine complex, already home to a major detector of neutrino particles. Together, the instruments would enable a much clearer idea of the location of a source to be confirmed.

Launch for LISA

LIGO has already achieved a string of detections, and as other detectors are added into the network it is likely to have many more significant successes. There is also a further phase of enhancement envisaged called LIGO A+, which is expected to take the detection rate of binary black holes to perhaps five a day by 2023. By 2028 LIGO Voyager, a major upgrade of the LIGO systems, could be installed, taking binary black hole detections up to several per hour. There has also been talk of future observatories in the 2030s such as Cosmic Explorer with perhaps 40-kilometre arms, capable of seeing pretty much every black hole merger throughout the universe with masses below 1,000 solar masses.

Nonetheless, LIGO-style ground-based observatories will always have their limitations. It was entirely possible that such instruments would never have detected anything, and long before LIGO succeeded in 2015, proposals were being floated to take a much bigger step, by moving gravity wave observatories into space.

Satellites make appealing homes for telescopes. Almost everyone has seen the stunning images from the Hubble Space Telescope. By comparison to its Earth-bound equivalents, Hubble is nothing special. Its mirror is 2.4 metres (95 inches) across. It's a relative baby compared even with the greatest telescope of the mid-twentieth century, Mount Palomar's 5.1-metre (200-inch) telescope. Meanwhile the biggest earthbound individual telescope of today, the Gran Telescopio Canarias, has a far larger 10.4-metre (409-inch) aperture. But the Hubble delivers for a good reason.

Moving your observatory into space takes your telescope away from the interference caused by sharing the Earth with humans and an atmosphere. For optical telescopes, the issues are mainly light pollution from cities and the scattering of photons by air molecules. Gravity wave telescopes, as we have seen, suffer particularly from the interference of vibrations caused by humans and their technology as well as by natural causes. In space, not only can no one hear you scream, they can't feel your vibrations either.

This means that an equivalent of LIGO operating in space would have immediate advantages in terms of the ability to operate without interruption, avoiding many of the false readings that otherwise have to be dealt with. But there's another opportunity that space has to offer. LIGO is limited in its sensitivity by the length of the arms of its interferometer. At 4 kilometres, they are impressive – but in space there would be no need for the evacuated tubes to run the laser beams down. Space is already a vacuum. And this means that the arms could be longer. Much longer.

The original concept for LISA (Laser Interferometer Space Antenna) dates back to the early 1990s, when it was hoped that it would have been live before 2010. LISA's interferometer arms would be provided by the gaps between three satellites positioned in a triangle with million-kilometre sides, giving far greater sensitivity than LIGO can provide. But, not entirely surprisingly, there was some reluctance to put large amounts of space exploration budget into a search when there was, at the time, no direct evidence of the existence of gravitational waves.

Unlike a ground-based observatory such as LIGO, LISA would have the chance to take in the whole of the sky. Rather than orbit the Earth as most satellites do, LISA is planned to be in an orbit around the Sun, following in the Earth's path at a distance of between 50 and 65 million kilometres, about 125 times the distance at which the Moon orbits. The hope is that LISA would operate for a minimum of four years, but it has been designed to be able to stretch this to ten years if all goes well. LISA's far longer interferometer 'arms' would enable it to deal with much lower-frequency gravitational waves than LIGO can detect, in the range between 0.001 and 0.1 Hz (ripples per second).

This would enable LISA to detect waves from much higher-mass black holes, to explore their role in galaxy formation and to use the interaction of these huge black holes with other bodies to find out more about the black holes' event horizons and to test out black hole theoretical physics. It would also be able to predict coming mergers of smaller black holes up to a week before they occurred, as initially the frequencies of an inspiral are low, so that ground-based detectors and light-based telescopes could be given a warning of an impending event. And LISA would be able to detect sources that were too weak to be picked up by an Earth-based system, such as the gravitational waves produced by white dwarf binary stars orbiting each other. The LISA proposal suggests that up to around 25,000 individual binaries could be identified and studied. And this is not to mention the inevitable hope that something new and totally unexpected will be found.

LISA was originally a joint venture between the European

Space Agency (ESA) and NASA, but in 2011, suffering severe funding restrictions, NASA pulled out. Initially, ESA looked likely to go for a scaled-down version, known as the New Gravitational Wave Observatory, but with a renewed interest in gravitational waves after the LIGO discoveries, in early 2017 a revamped version of LISA, now featuring 2.5-million-kilometre beams, was proposed and at the time of writing has just been accepted for funding. This followed the test launch in 2015 of the LISA Pathfinder, a single satellite with tiny 38-centimetre (15-inch) interferometer arms, which has already exceeded expectations in its capabilities.

The 2015 observation at LIGO and the subsequent detection events have arguably changed everything. Where the original proposal for LISA was very much fumbling in the dark, with the distinct probability to be faced that gravitational waves would never be detected, we now know that detections are possible and that events detectable by the much less sensitive LIGO are happening quite frequently. With LISA's capabilities, gravitational wave astronomy could truly come of age.

It's not going to happen quickly. The LISA satellites are currently due to launch in 2034, with another year required to get LISA active – and slippages are not uncommon. But there is far greater determination now that gravitational wave detection is known to be practical.

Gravitational channels

In the future, there are potentially four distinct bands available to gravitational wave detection observatories, much in the same way as we have light telescopes that work with the distinct frequency ranges of radio, infrared, visible, ultra-violet, X-rays and gamma rays. Similarly, the bands in gravitational wave astronomy are based on the frequency range of the waves. LIGO concentrates on those with frequencies around 10 Hertz – each cycle of the wave lasts between around 100 milliseconds and 1 millisecond. LISA would detect gravitational waves with periods of minutes to hours, produced by merging massive black holes and a wider range of sources.

An alternative approach to detection is called pulsar timing arrays, where radio astronomers monitor the frequencies produced by the spins of pulsars. As a gravitational wave passes through the Earth it should very slightly speed up and slow down the rate at which the pulsars appear to be 'ticking', so the pattern of changes as seen from an array of small radio telescopes can be used to track the wave. That approach would cope with gravitational waves with periods of years to decades.

The final possibility is the technique that was attempted in BICEP2 (see page 106): detecting waves with periods of millions to billions of years as a result of polarisation changes in the cosmic microwave background, if this technique can be used effectively.

To carry these detections forward is going to need significantly more investment. Which inevitably leads to the question – how can we justify spending such large amounts on single projects? What is the point of so-called 'big science'?

What's big science for?

Whenever scientists propose huge projects such as LIGO, there is a question mark over the spending involved. If a project costs a billion dollars or more, it isn't unreasonable for the taxpayer who foots the bill to ask 'What am I getting for my money?' and, perhaps more importantly, 'Could all that money be spent on something more useful?'

When cash is tight, there are always challenges to blue-sky research. Administrations often attempt to focus spending on scientific work that has a specific practical outcome in mind. However, the problem with this approach is that it's very rare that we know in advance what the practical implications of a pure research project are going to be.

Think, for example, of quantum physics. This involves apparently obscure, abstract theory about the bizarre behaviour of tiny particles such as electrons and atoms and photons of light. When quantum theory was being developed in the early decades of the twentieth century, no one asked 'What is it for?' It was work on the fundamental behaviour of the essential components of reality. As it happened, there was a huge benefit that arose from this work. Quantum theory enabled the development of modern, solid state electronics. It has been estimated that 35 per cent of GDP in developed countries is dependent on quantum technology, primarily in the form of electronics. But no one knew that this would be the outcome of that early research.

In those early days of quantum physics, big science was virtually unheard of. Arguably, the first true big science project was the Manhattan Project to develop nuclear weapons during the Second World War. This certainly was driven by specific goals. However, since then, a number of extremely expensive projects, such as the Large Hadron Collider, LIGO and various space science developments have pushed the boundaries of scientific expenditure. Often these endeavours are about discovering fundamental new ways of looking at the universe.

It is arguable that, as with the arts, we ought to justify this kind of fundamental research because it's part of what makes us human. We have a passion to know more about how the universe works, and we ought to be putting effort and money into supporting that drive. A parallel can be drawn with the old story that during the Second World War, Winston Churchill was asked to reduce funding for the arts in order to support the war effort, and said in reply: 'Then what are we fighting for?' Sadly, the story itself appears to be a myth, perpetuated by an internet meme. Churchill certainly considered the arts essential and said so before the war, but he appears never to have made this particular statement. However, this doesn't make the concept any less valid.

Assuming, then, that we should be doing fundamental research with no specific application in mind, there is also the question of where the limited resources available should go. When the American Superconducting Super Collider (SSC) was scrapped, making Barry Barish available to run LIGO, it was because the funds went instead to the International Space Station. The long-term result has certainly been bad for science. Instead of spending significantly less than the Large Hadron Collider cost for an arguably better piece of equipment, ten times the cost of the SSC has now been spent on the Space Station with no significant scientific outcomes whatsoever.

Was this a bad choice? Yes and no. It was a bad use of the science budget, but it's arguable that space exploration is also an essential for human survival and growth – just not as a competitor to science. It's not really about science. As Stephen Hawking has said: 'We are running out of space and the only places to go to are other worlds. It is time to explore other solar systems. Spreading out may be the only thing that saves us from ourselves. I am convinced that humans need to leave Earth.'

It would arguably be more appropriate if space travel were funded from the defence budget rather than competing with science. Meanwhile, a handful of major science projects can surely be justified where they are making big inroads into our understanding of the universe and our place in it. LIGO was a huge gamble. But it has paid off. And we should remember that as well as transforming the future of astronomy it has produced a collaboration between sixteen nations, enabling over 1,000 people to make a contribution. Most of all, though, it provides that step forward in astronomy – the next big step since the introduction of the telescope.

Astronomy's latest

Conventional light-based astronomy hits a barrier if we look far enough out into space. As we have seen, space is a kind of visual time machine. The further you look, the further back in time you are seeing, as light takes time to reach us. So, for instance, when we look at the nearest star to us other than the Sun, Proxima Centauri, we see it as it was about four years ago, while we see the nearest major neighbouring galaxy to our own, the Andromeda Galaxy, as it was around 2.5 million years ago. However, when we get back to a point around 380,000 years after the Big Bang, we can see no further.

This is because the universe was a cloud of opaque plasma before that point, so no light could pass through it.

The cosmic microwave background (see page 107) is the light that first started travelling when the universe became transparent. But gravitational waves have no barrier to their movement beyond that point and by using them we should be able to see back close to the Big Bang itself. If the theoretical inflation process that it was hoped BICEP2 would provide evidence for (see page 108) really happened, then it's unlikely gravitational waves would have survived from before that, as the sudden immense stretching of space would iron them out – but we should be able to see anything that came after.

That's not to suggest we would necessarily be able to say nothing about waves from the Big Bang, as it's these 'primordial gravitational waves' that are likely to have contributed to the polarisation process in the cosmic microwave background that BICEP2 was searching for. But we would then be back to indirect observation, rather than true gravitational wave astronomy.

Another theory that could be explored with the penetrating power of gravitational waves is the splitting of the electroweak force into electromagnetism and the weak nuclear force.* It is thought that these two forces were originally unified, but after the inflationary period they split in a phase transition process known as spontaneous symmetry breaking. If this were the case, and the phase transition happened in a particular way, expanding and colliding 'bubbles'

---

* Electromagnetism and the weak nuclear force are two of the four 'fundamental forces' of nature (the other two being the strong nuclear force and gravity). Electromagnetism is responsible for the interactions between matter and matter, plus matter and light, other than gravity, while the weak nuclear force is involved in the conversion of particles from one type to another occurring in nuclear reactions.

of space could have generated their own distinctive gravitational waves which should now be in the kind of frequency range that the LISA observatory could deal with.

For that matter, there are even exotic alternative theories to the Big Bang such as the 'bouncing branes' ekpyrotic universe concept, which sees our universe as a three-dimensional membrane or 'brane' in a higher dimension which is stretched out as it expands, losing 'crinkles' that in the process would produce distinctive gravitational waves – so this new type of astronomy could even help make it clearer exactly how the universe originated.

The ekpyrotic model of the universe is based on M-theory, the souped-up version of string theory, one of the approaches to providing a common theory that unifies the quantised forces of nature, electromagnetism and the strong and weak nuclear forces, with the very different force of gravity. String theory requires nine spatial dimensions and M-theory needs ten. The extra dimensions we don't see are supposed to be curled up so tightly that we don't notice them. However, it's possible that if a theory such as this, requiring extra dimensions, is correct, gravitational waves will enable us to detect the indirect effects of these extra dimensions.

By modelling the impact of curled-up extra dimensions on gravitational waves, scientists at the Max Planck Institute in Germany believe that if the dimensions are present, they would generate a series of unusually high-frequency overtones in the gravitational waves that wouldn't otherwise be seen. They should also subtly change the way that space expands and contracts as gravitational waves pass through. Although such observations are unlikely any time soon, it's another potential application of this new way of looking at the universe.

Light in the darkness

There is one other major area of astronomy where gravitational wave technology could prove crucial – dark matter. There should be five times as much of this hypothetical substance in the universe as ordinary matter, but astronomers can't see it because, if it exists, it appears not to interact with light or other matter via electromagnetism. You can't see it; you can't touch it. But it makes its presence felt gravitationally – so what better target for gravitational wave astronomy?

Back in the 1930s, Swiss astronomer Fritz Zwicky noticed something odd about a group of galaxies called the Coma Cluster. Like most things in the universe they were spinning round, and when things spin too fast, they fly apart. What holds galaxies together is gravity. But even with all the stuff in these galaxies, they were spinning too quickly to stay together. They should break up like clay on an over-enthusiastic potter's wheel. The implication was that there was extra matter that we couldn't see. Zwicky was largely ignored, but in the 1970s, when we had a better idea of what was out there, the American astronomer Vera Rubin noticed similar oddities in spiral galaxies like our Milky Way.

Even with everything that we know makes up a galaxy, there's not enough for them to hold together. Zwicky imagined there must be another kind of stuff, invisible stuff, which we now call dark matter. For a long time, physicists have been searching for a particle that acts like dark matter – but none has been found.

It is arguably odd that they've mostly looked for one kind of particle – as normal stuff is made up of quite a few different particles, and there's no reason why dark matter can't be as well. It has even been hypothesised that there could be a whole dark matter universe with dark suns pumping out dark light. But as yet this is all science fiction, particularly as there is increasing suspicion that dark matter may not exist at all.

Dark matter is one possibility to explain what's happening – but not the only one. It's possible that ordinary matter behaves a little differently in the vast collections of galaxies than in ordinary objects. And if it did, it would explain most of the oddities we need dark matter for. It has also been suggested that the whole dark matter business is an error in the calculation of how much ordinary stuff is out there, which inevitably is an approximation. The jury's out on this one.

Getting a handle on dark matter is essential if we're going to understand what's going on in the universe – and like quantum physics, who knows whether there might be practical implications of understanding it in the future? Gravitational wave astronomy has the potential to give us the tool to further explore the nature (or absence) of dark matter. By definition, this is a part of the universe for which light-based astronomy is useless. But for gravitational wave astronomers, dark matter is just as real and present as any other kind of matter.

If the right sources can be found that involve large volumes of dark matter, moving in such a way that waves are produced, then we have a real possibility of making clear distinctions between what would be produced by an accumulation of dark matter and what would be produced by variants in gravitational theory. As yet, we can't make the call. Some observations are more supportive of dark matter, others of modified Newtonian dynamics or misinterpreted statistical data. However, it's possible that the gravitational wave observatories of the future will pin down exactly what is happening to make it seem that there is far more to the universe than we have so far been able to detect.

Whether or not dark matter is finally observed, though, there is no doubt that the development of gravitational wave astronomy is a remarkable confirmation of human capabilities when faced with apparently impossibly difficult odds.

A human triumph

The success of advanced LIGO is a genuine triumph of human ingenuity and staying power. For decades, gravitational wave scientists persisted in building detectors that found nothing – yet this did not stop the LIGO consortium from taking a leap in the dark and, perhaps more remarkably, persuading funding bodies to come up with the cash.

When LIGO finally got to the stage of having a reasonable chance of detection in 2015, it was dealing with ridiculously small movements in a system that had to ignore far greater impacts from the environment. The Large Hadron Collider might be the biggest experiment in the world, but it's hard not to see LIGO as the most sophisticated. There was luck involved in that first event being so clear – yet this was combined with the dedication of those 1,200 people working both directly on the design and operation of the detectors and across the globe in analysing the data and developing the theoretical templates against which the signal would be tested.

We should also remember the driving force and creativity of the three key physicists behind LIGO – Rainer Weiss, Kip Thorne and Ron Drever. This was rightly recognised in the announcement in October 2017 that the Nobel Prize in

Physics would go to the detection of gravitational waves. Drever sadly died in March of that year, and the prize cannot be awarded posthumously, but Barry Barish was rightly included for his role in turning the project around.

The kind of research done with LIGO may never produce a major breakthrough that will have an everyday use – but it will do much to improve our understanding of the universe we live in. It will contribute greatly to the sum of human knowledge. Though this kind of fundamental exploration of the nature of the universe may not have practical applications for our daily lives, it is surely an example of what makes human life more than simply the struggle to survive from generation to generation.

By detecting gravitational waves and pushing back the boundaries of our understanding, we confirm the strength of the human spirit.

FURTHER READING

Waves
Demonstrations of transverse and longitudinal waves
are available on the *Universe Inside You* website at
universeinsideyou.com/experiment5.html

Light
Background on our exploration of the nature of light and the
wave/particle debate – *Light Years*, Brian Clegg (Icon Books,
2015)

Gravity
General overview of gravity – *Gravity*, Brian Clegg (St Martin's
Press, 2012)
Einstein's general theory of relativity – *Einstein's Masterwork*, John
Gribbin (Icon Books, 2015)
Relativity, including gravity's place in it – *The Reality Frame*, Brian
Clegg (Icon Books, 2017)
History of our understanding of gravity and the search for
quantum gravity – *The Ascent of Gravity*, Marcus Chown
(Weidenfeld & Nicolson, 2017)

Relationship between black holes and the rest of the universe
through gravity – *Gravity's Engines*, Caleb Scharf (Allen Lane,
2012)

Gravitational waves

Demonstration of the basic types of wave – universeinsideyou
.com/experiment5.html

Overview of the search for gravitational waves and the theory
behind it – *Discovering Gravitational Waves*, John Gribbin
(Kindle Single, 2017)

Overview of the development of interferometer gravitational
wave observatories – *Black Hole Blues*, Janna Levin (Vintage,
2017)

Sociology of gravitational wave discovery

Day-by-day diary of the discovery process for the September 2015
event – *Gravity's Kiss*, Harry Collins (MIT Press, 2017)

Study of the two blind injection events – *Gravity's Ghost and Big
Dog*, Harry Collins (University of Chicago Press, 2014)

Exploration of the early years of gravitational wave research –
Gravity's Shadow, Harry Collins (University of Chicago Press,
2004)

INDEX